On the Verge

Rebecca D. Costa
author of *The Watchman's Rattle*

RosettaBooks®

NEW YORK 2017

For information, please contact RosettaBooks at production@
rosettabooks.com, or by mail at One Exchange Plaza, Suite 2002, 55
Broadway, New York, NY 10006.

RosettaBooks editions are available to the trade through Ingram
distribution services, ipage.ingramcontent.com or (844) 749-4857.
For special orders, catalogues, events, or other information, please write
to production@rosettabooks.com.

First edition published 2017 by RosettaBooks

Cover art by RichVintage/Vetta/Getty Images
Cover design by Christian Fuenfhausen and Brehanna Ramirez
Interior design by Brehanna Ramirez

Library of Congress Control Number: 2017940684
ISBN-13 (print): 978-0-7953-5057-3
ISBN-13 (epub): 978-0-7953-5060-3

www.RosettaBooks.com
Printed in the United States of America

RosettaBooks®

CONTENTS

For Edward O. Wilson

PREFACE

Ask anyone on the street "What's the single most important attribute—the one above all others—vital to success?" The first answer you'll get is money or power. A few may say persistence or a good education. And, some will throw in family and strong role models for good measure.

But is this really true?

Not from a sociobiologist's perspective.

For it is not the wealthiest, strongest, fastest, or most cunning—not the most educated or experienced—which prevail when the environment takes a turn. More than a hundred fifty years ago, Charles Darwin revealed that it is *our ability to adapt* that is the determining factor. And while that may not sound like news today, when you stop and think about it, our attitudes toward adaptation haven't changed in more than a century. We continue to treat adaptation the way nature does—as some random roll-of-the-dice over which we have little or no control. When the environment changes, some individuals and groups thrive, some hang on for dear life, and others perish. And that's that.

In truth, we understand very little about how to make ourselves, our workplace, our economy or government more "adaptable." For example—is there any *one* skill that matters more than others when it comes to responding more quickly, more precisely, more successfully to change?

It turns out there is.

For it is the organism with the greatest foresight that has the upper hand in any situation.

Foresight allows us to make plans, avert danger, get the jump ahead of others. Foresight commands us to fashion the spear before the attack, sell a stock before it tumbles, cut a cancer before it spreads. Foresight is the precursor to opportunity, the warning before the storm, the salvation before the sin.

There can be no greater advantage than certainty of the future.

And when it comes to man's prophetic powers, this book could not arrive at a better time. Recent breakthroughs in technology—such as the proliferation of predictive analytics, Big Data, and sensor and satellite technologies—have made it possible to anticipate future outcomes with unprecedented accuracy. With every passing moment our forecasts grow more robust, more consequential, more intrusive. That's because foreknowledge has turned previous notions about adaptation on its head. We are no longer adapting to changes in the environment. We stand on the cusp of *changing the environment to which we must adapt.*

And by environment, I mean every environment: physical, economic, political, social... every aspect of human life is affected by our ability to foretell future consequences—and the growing burden to avert antagonistic events before they occur.

Darwin's world is no more. Even a stalwart theory like evolution must yield to changing circumstance. And while it may seem sacrilegious for a sociobiologist to claim science and technology have brought humankind to the brink of transcending laws that have governed life for billions of years, the truth has no obligation to conform with previous precedent. The world has changed. Humans have changed it. And we must now come to terms with the deeper meaning of that change.

On the Verge is the story of tilting the odds in our favor, manipulating outcomes before the fact, and preempting failure. It is the story of how we rose to become the most elastic organism on Earth and, along the way, unlocked the secret to eternal prosperity. It is the story of where we have been, and where we are headed. It is the story of foresight: the crowning achievement of human ambition.

Rebecca D. Costa
November 10, 2016
Big Sur, California

"Want of foresight,
unwillingness to act when action
would be simple and effective,
lack of clear thinking,
confusion of counsel until the emergency comes,
until self-preservation strikes its jarring gong,
these are the features which constitute
the endless repetition of history."

- Winston Churchill

CHAPTER ONE

Foresight

Imagine a world where a handful of businesses and governments could foresee the future. In the beginning, they could see to the end of the block. A little later, the end of the street. Then all at once, around the corner, down the highway, across the ocean, and beyond the curvature of the Earth.

Now imagine for a moment their prognostications were right.

Not right once or twice. Right every single time. Imagine anticipating future events with such precision that a threat could be quashed before it had opportunity to materialize; shortages and surpluses could be managed in advance; public opinion shaped beforehand.

Well, imagine no more.

This phenomenon is underway—a shift so subtle it feels like no more than a hand gently ushering us across a crowded room. That crowded room is the Information Age—and we are slowly making our way to the other side to *a future that is knowable.*

A knowable future? Have I lost my mind?

Perhaps.

But consider the evidence. Fifty years ago, the sex of a newborn was a surprise revealed only at birth. A few decades ago, we didn't know whether a person was predisposed to breast cancer, baldness, Alzheimer's, or depression. We didn't have the meteorological models, instrumentation, or satellite imagery to evacuate an entire city in advance of a hurricane. And no way to know when a country's currency was on the verge of collapse. Never mind how oil production in the Middle East will affect banana prices in Tokyo...

Every day our ability to anticipate future outcomes grows more acute, more all-encompassing, and extends further out. This sea change has equipped today's leaders with a previously unimaginable power—the power to respond to and shape events before they occur. We stand on the cusp of what Darwin himself might have called *predaptation: the ability to adapt a priori.*

Similar to other leaps in human evolution, this capability did not occur overnight. Following millions of years of trial and error—and the failure of over 99 percent of the species that once inhabited Earth—Mother Nature saw fit to bless a single organism with the aptitude to understand "tomorrow." Our adeptness at conducting sophisticated thought experiments—our ability to assess risk and prioritize complex scenarios in rapid fire—is unique to *Homo sapiens.* Not only is this faculty unequalled, it is also, without question, the most powerful asset nature has produced to date. For there is no greater advantage than the ability to abate danger or seize opportunity which has yet to manifest. Not in nature. Not in business. Not in governance.

The truth is, every controversy, every intractable problem we face today, every emotionally charged debate is about events we see coming: how to stop climate change; how to save the Euro; how to quash the threat of nuclear war, illegal immigration, and terrorism; how to care for a growing aging population; how to reverse debt, obesity, gun violence, racism, over-fishing, and addiction. We see these things with new foreboding—though we passionately disagree on what to do about them. Or whether anything *can* be done.

It is fair to say that our initial foray into the future has been clumsy—as awkward as our first attempts at two-legged loco-motion. Despite possessing more knowledge and technology than at any other time in human history, we behave schizo-phrenically toward the future, vacillating between a destiny that is shaped by free will, one that is erratic, indiscernible, and uncontrollable, and one micromanaged by supernatural forces. When a person carelessly invests their money and loses it, when they eat their way to obesity or illness, when they commit a crime and are sentenced, we view these events as the predictable consequences of our actions. But let those same individuals fall in love or get a flat tire on the road and we treat these events as the work of Gods who shuffle the deck for their own pleasure.

Well, which is it? *Can we control the future or not?*

The ancient Greeks said no. They created three goddess-like "Fates" called the Moirai to explain the trials and triumphs mortals encounter. The Moirai controlled every aspect of human life from birth through death. As the ordained enfor-cers of destiny, even Gods like Apollo and Poseidon were rendered impotent by the Moirai, let alone the feeble ambitions of men.

Throughout Asian cultures, the future is treated as a time when justice is administered through reward and retribution. According to the principles of Karma, every living organism is held accountable for their actions. A mindful life leads to health, safety, peace, and wealth. Whereas an unconscious life—one that inflicts harm on others—produces suffering for the offender. This is how balance is maintained in the universe.

In areas of Africa and South America, sacrifice plays a large role in determining the future. Small animals, money, food, song, and dance are offered to deities to court favor. Whether that favor is a robust harvest or the birth of a healthy child, villagers come together to perform rituals and offer gifts aimed at turning the tides of fortune.

No matter which culture, religion, or period in human history we examine—from Christianity and Islam to Santeria, from European high society to remote tribes in New Guinea, from the great Egyptian, Ming, and Roman empires to the twenty-first century—the future is portrayed as a tug-of-war between man, chance, and God.

It wasn't until recently this ambivalence was brought to heel. For the first time, predictive algorithms powered by lightning-fast computers and mobile communications brought the entire universe of human knowledge to man's fingertips. Technology made it possible to string together millions of variables, in real time, revealing cause-and-effect relationships we never knew existed. And this leap has armed humanity with a staggering power: the power to reverse-engineer the consequences of our most benign actions.

With the Information Age came data. With data, analytics. With analytics, foreknowledge. And with foreknowledge, *foresight*. As our prowess for prognostication spread, we stumbled upon an unexpected truth—one which has tectonic implications: *there is far less randomness to the future than we thought. More of it is predictable than not.* And if more is predictable, then more can be manipulated.

And overnight the race to engineer the future was on.

China pulled ahead of the pack, sewing up rare earth minerals throughout the African continent—exporting over a million Chinese citizens to settle there. The United States and Russia staked an early claim to outer space. Visionaries like Richard Branson, Elon Musk, Craig Venter, and Ray Kurzweil got a head start in commercial space vehicles, artificial intelligence, and genetic engineering. Hundreds of start-ups jumped into sensor and satellite technologies designed to measure anything, anytime, anywhere. Even large retailers got in on the action, quickly locking in milk supplies and prices when rises in temperature were forecasted. Once a connection between warm weather and a decrease in a cow's milk production could be established, grocery chains began monitoring NASA's meteorological database to corner supplies before shortages occurred.

As you read this page, businesses and governments are preparing for drones to deliver everything from life-saving medicines to common parcels within minutes.

Last year, Amazon filed for a patent on a blimp-like "floating warehouse," disrupting previous notions of fixed, terrestrial storage facilities, distribution centers, and factories.

Today, forward-looking insurance carriers and automobile manufacturers are making plans for self-driving cars—vehicles that will eliminate millions of accidents caused by operator error, along with the need for driver training, driver's licenses, law enforcement, stop signs, and traffic signals. Automobile ownership will become a thing of the past, and there will be no more need for taxis. Healthcare and pharmaceutical companies are preparing for nanobots—smaller than a single cell—that will treat disease from the inside out, making surgery and drug-based treatments obsolete. Retailers, immigration departments, and educators are getting the jump on facial recognition software to screen applicants, identify criminal conduct, and track student and consumer reactions. And political pollsters are not far behind. The only people not surprised by the 2016 election of Donald Trump were technologists who had better tools and data.

We can foresee that within the next decade every human will confer with an electronic cyber twin that looks and behaves the way they do—one which has the ability to quickly search the Internet and deliver only the data their biological counterpart needs to make a decision. Our electronic avatars will know our preferences, our habits, our history, and will allow us to accomplish twice the work in half the time. And bio-feedback sensors will instantly alert us to nutritional deficiencies, dehydration, and the earliest indications of disease.

Every nation will rely on an electronic army to fend off cyber attacks and defend their physical borders; commerce will be driven by a single, universally accepted cyber currency that will eliminate the power to artificially manipulate exchange values; all energy will be renewable and free as breakthroughs in renewable sources, battery storage, etc., reach previously unimaginable capacities; sophisticated 3-D printers will print food to fit our palette preferences, as well as perfectly fitted clothing, furniture, and other goods from the comfort of our homes. Hospitals will use these same 3-D printers to produce prosthetic body parts on the fly, architects will print flawless buildings with the electrical, plumbing, and infrastructure in place, and auto parts companies will print spare parts on demand. This technology and robotics will deal a final death blow to volume manufacturing, causing intellectual property to become the only valuable corporate asset while simultaneously giving rise to a class of "knowledge workers," the likes of which the world has never known.

These changes are coming. They will be upon us soon.

Ready or not.

"When one admits that nothing is certain
one must, I think, also admit
that some things are much
more nearly certain than others."

- Bertrand Russell

CHAPTER TWO

To See Or Not to See

The day I turned sixteen I began looking for my own wheels.

I didn't have much money—a little babysitting and typing on weekends was the only work available. One summer a neighbor paid me to sleep at their house and feed the dog while they went on vacation. I also counted on my grandmother to send me a little something on birthdays and holidays. It wasn't a lot, but when all you have are nickels, dollars feel big.

One morning I was sitting at the kitchen table studying my options in the classifieds when my dad walked in.

"What're you looking for?

"A car," I said without looking up.

He poured himself a drink and sat down. "What kind of car?"

"A Bug."

He shuffled through the rest of the paper and pretended to scan the sports page. But I knew what he was doing...

"Anyone going with you?" he piped up.

"Nope." I was sixteen. I wanted to get my own car.

For a long while neither of us spoke. He with his section of the paper, me with mine. Then he cleared his throat.

"Well... if you want my advice... maybe you should take someone with you. Someone who knows something about cars..."

It was an offer. I kept on reading.

"Okay, then." He stood up, grabbed his coat, and headed for the garage.

Then he stopped. "It's your money. And I can see you don't want any help. But when you get out there, look on the ground for oil. And if the car has more than one hundred thousand miles on it, remember you're just buying someone else's problems." Then he disappeared out the door.

Someone else's problems?

In 1970, a car with one hundred thousand miles spelled trouble. Today, the odometer on my Land Cruiser reads 276,000 miles, and—knock on wood—it hasn't required a single major repair. But since I was conditioned by my father to expect the engine to fall out after a hundred thousand miles, I recently asked my dealer to give it a once over. They did. Outside of a sticky antenna and a small tear in the passenger seat, there were no impending signs of doom. In fact, the mechanic who inspected the vehicle said, "You could go four, five hundred thousand miles before you see a problem with this car."

No problem for half a million miles?

It's mind-boggling to think the lifespan of an automobile has jumped fivefold since my first license.

How did that happen? More to the point—*how did the mechanic know how long my car would run?*

The answer is technology.

Human ambition has never shined more brightly than when it comes to advancements in science and technology. In addition to possessing more knowledge about our physical universe than ever before, we've developed tools to amplify our ability to foresee failure—tools capable of spotting and shoring up weaknesses long before they become a problem. So whether it's the transmission, power steering, or heated seats, engineering foresight has produced cars that now run trouble-free for a half a million miles.

And not just cars. Everything can be made safer, cheaper, faster, and more dependable than before. It's no wonder lifetime warrantees have become commonplace, and once-familiar television and appliance repair shops have disappeared. Everyone knows it's cheaper to replace a broken appliance than try to fix it. Assuming, of course, you can find the parts.

So what was the breakthrough that allowed us to get out in front of product defects in one generation? Perfect goods in ways we previously couldn't? What propelled engineering from fixing to foreseeing?

Predictive and Preemptive

Though the first computers were designed to tabulate figures faster than any team of humans could, it didn't take long before we discovered these machines were also better at aggregating, comparing, contrasting, and manipulating figures. Soon, we began gleaning powerful new insights about the data we were amassing. Then, as computers moved out of hermetically sealed facilities onto our desks and into our homes, programs designed to spot errors began cropping up everywhere. Misspell a word and the computer would underline it. Add the figures incorrectly on a spreadsheet and the miscalculation would change color. Miss an entry and we were prohibited from moving to the next screen.

While these features had a tremendous impact on quality, simply pointing out our spelling, math, or engineering errors was far from optimal. That's because they worked *after* an error had already been committed. In other words, at best, they shortened the time between making, finding, and correcting an error—often compressing the timeframe to seconds.

But still, it was *after the fact*.

Then came the next phase in computing. Overnight the Internet and mobile communications made it possible to access the sum of human knowledge anywhere, anytime. And just as quickly, smart machines aimed at *preventing* errors came on the scene. In the blink of an eye, we transcended from reversing mistakes to circumventing them. And it was this paradigm shift that gave new utility to foresight.

Before Is Better Than After

In the late '70s, I had the good fortune to work for one of the companies responsible for this transformation. Headquartered in California's Silicon Valley, Calma Company was surrounded on all sides by companies that were changing the way humans would one day work and live. Apple was busy putting a computer in every home, Ungermann-Bass was building the first enterprise-wide network, Omex was pioneering optical storage, and Motorola was developing the first cellular phones. It was a time when the best and brightest came pouring in from China, India, Korea, Japan, Russia, and every town in America. At one point there were more PhDs and engineers within thirty square miles than could be found in entire countries.

Though technology was exploding on all fronts, the impetus behind the westward stampede was the semiconductor and integrated circuit. Integrated circuits did exactly what their name implies: they were *integrated*—a collection of millions of interconnected transistors, capacitors, resistors, and other components reduced to less than one-hundredth the width of a human hair. Though no one realized it at the time, these miniature components were destined to become part of everything powered by electricity—from massive factory machines and spacecraft to coffeemakers, wristwatches, and children's toys.

But there was one small problem...

Throughout the '60s and '70s, the engineering drawings for integrated circuits consisted of an incomprehensible maze of dense lines and symbols plotted on sheets of paper the size of a dining table. These manually drawn designs were not only multilayered, but extremely intricate. So intricate that engineers often resorted to printing every layer of a circuit on individual sheets of velum, then carefully laying one piece of velum on top of the other to see whether the components and electrical pathways on one layer interfered with activity on other layers. Imagine, for a moment, dozens of sheets of clear plastic with convoluted mazes stacked one on top of the other. Now imagine trying to spot problems *between* the sheets of plastic.

Impossible!

So, for a brief period of time, it looked as though integrated circuit design had reached the limits of human ability.

The founders of Calma saw the standoff coming. They were the first to convert the design rules used by engineers into software. They married that software to an electronic drawing tablet, minicomputer, storage device, and flatbed printer, and presto! The world's first computer-aided design (CAD) system for electronic circuit design was born.

The first CAD stations worked a lot like an electronic "Etch-A-Sketch"—the red plastic toy children use to draw vertical and horizontal lines by manipulating two dials. Only this Etch-A-Sketch was smarter than its operator. It came equipped with everything known about circuit design. Every component. Every specification. Every design rule. Every nuance. Want to find the shortest path to connect two components? The optimal route would light up on a graphics screen. Want to know how fast the finished circuit would perform? The CAD system would simulate performance and point out areas where speed could be improved. Want to know how much the circuit would cost? An itemized Bill of Materials would appear. With the advent of CAD, billions of design decisions were relegated to a machine—one capable of optimizing performance, reliability, and cost better than the most experienced engineer.

CAD was one of many advancements responsible for the birth of *automated preemption*: technology designed to head off problems *before the fact*. No more trial and error. No more guesswork. No more endless testing of physical prototypes, tedious iterations, or messy postmortems to identify the cause of failure. Everything about the circuit was known before it ever reached the manufacturing floor.

While Calma was changing preemptive circuit design, other CAD companies such as Auto-Trol, Applicon, CADAM, Computervision, and Intergraph were developing similar tools for other industries. Whether a product was a nuclear submarine or skyscraper, computer-aided design, computer-aided engineering (CAE), and computer-aided manufacturing offered businesses a better way to design, test, build, and foresee future outcomes.

That new dishwasher, coffeemaker, television, camera, and car you just purchased? Today's manufacturers know exactly how many loads of dishes it will clean, cups of coffee it will brew, photographs it will snap, and miles it will travel before problems will occur. They know which components will fail first, second, and third, and how they will fail. And it is this foreknowledge that instructs today's businesses on which parts to warrantee, which to mark up, which to stock and train dealerships to replace—even when to start sending enticing upgrade offers, and to whom.

Engineering foresight has come a long way.

I don't mean to suggest that our ability to foresee failure has become so bulletproof that there aren't moments when we're caught off guard. When levees give way to floods, when airbags, children's toys, and medical implants malfunction, when we ignore the thermal effects that cause a space shuttle's heat shields to fail, the consequences are grave. But within a short period of time, we get to the bottom of these problems.

Then—confident we've identified the culprit—we return to outer space and get busy redesigning levees, equipment, and buildings. Only this time, better.

Last week, my ten-year-old toaster quit. I put bread in, the bell rang, and out popped bread again—no toast. For a moment I considered taking it apart to see what was wrong. Then I began thinking about the lifespan of a new toaster—one perfected by today's CAD systems—one that takes advantage of the new state-of-the-art engineering and manufacturing techniques—it occurred to me I will need only one more toaster for the rest of my life. It was a sobering realization. So I went out and bought the best one I could: the toaster the Queen of England uses. Now, barring a house fire, I'm done with toasters. According to the warranty, I will reach my *Mean Time to Failure* before my new appliance does.

Reverse-Engineered World

Once I became aware of just how powerful an advantage foresight is, I let it take over my life. These days I reverse-engineer everything. I start with the outcome I want (or don't want) and work backward. And it's not just me. Everywhere I look reverse-engineering has taken over. Students pick their colleges according to the job they want when they graduate. Politicians use polling data to shape their campaign messages and platforms. Carbon credits are reverse-engineered based on their future impact on our environment. The entire insurance industry is based on reverse engineering—eligibility and premiums are determined by future risk. What do we think retirement planning and drafting a will or prenup are about?

Take something as straightforward as obtaining a mortgage for a new home.

One of the primary barometers lenders use to qualify borrowers is a mysterious algorithm called a FICO score (Fair Isaac Corporation) which is designed to gauge a person's creditworthiness. So, if we want to be sure our mortgage is going to be approved, it behooves us to check our FICO score in advance of applying. Then head off anything negative that may impact our score.

So that is what I did.

Once I made the decision to buy a small cabin in Oregon where I could think and write, I quickly went online to check my credit score. Since I pay my bills on time, owe little debt, and have used the same credit cards for thirty years, I expected my score to be nearly perfect. But, to my surprise, the three big credit bureaus didn't agree. They considered me only a "good" risk, not an "excellent" one. And I had no idea why.

Now, if I had made this discovery a decade ago, I would have telephoned the credit bureaus, waited for them to each mail me printed reports, then tried to make heads or tails out of what they sent me. Even then, I might not have been able to get to the bottom of what was affecting my score. But today, thanks to a suite of easy online tools, there, right next to my number, was a breakdown of the factors negatively affecting it. And next to that, tips on what I could do to raise the number and by exactly *how much!*

In my case the culprit was my ratio of debt to credit. I may have owed little, but what I owed represented more than 30 percent of my available credit—which wasn't much since I pay in cash, and haven't applied for a new card since Reagan was in office. But apparently, 30 percent is some threshold.

So—using the powers of foresight and reverse engineering—I could see that I had three options: a) pay down what I owed so it was under 30 percent, or, b) open up more cards to increase my overall credit, or, c) ask for the limits on my current accounts to be raised. In other words, make my debt smaller or credit bigger. So, to be on the safe side, I did both. And presto! My score shot up nineteen points. The exact amount the web site predicted.

My point is simple. By acting to alter my FICO score, I have changed my future. I can now predict, with great certainty, that, barring an act of God, my mortgage will be approved. And I can also predict that I will qualify for the best interest rate available. The cabin is mine.

The Verge

What does it mean to stand on the verge of knowing future outcomes? To move from projecting to prescience? To allow data analytics to guide policy and action rather than tradition, politics, and the vestiges of prehistoric emotions?

It means we are in transition.

It doesn't matter if we're talking climate change, terrorism, government debt, or pandemic viruses, the challenge is the same: the reticence to move forward on our growing arsenal of foreknowledge.

Thankfully, help is on the way. Technology designed to join the past, present, and future into one seamless algorithm. Systems that treat every new piece of data as part of a larger, more complex pattern. Machines which know no ambivalence, prejudice, or limitations.

And they arrive none too soon.

"…forecasting, irrespective of its failures,
will never be abandoned. It is an inbred necessity
of human nature. The more we can anticipate
the course of events in the world in which we live,
the better prepared we are to react
to those events in a manner that can improve our lives."

- Alan Greenspan
The Map and the Territory

CHAPTER THREE

Jumping the Jar

It was common for me to finish my lab work early and this day was no exception. While other students waited to have their dissected fetal pigs examined by the teacher, I got a head start cleaning my station. I carried my lab notes (which, in sixth grade, consisted of a crude outline of a pig with handwritten call-outs) to the teacher's desk. But when I reached the front of the room, I heard a faint pinging sound. I stopped. Ping. Pa-ping. Ping. Ping. It seemed to be coming from an empty jar behind the desk.

I walked around, picked up the jar, and there, at the very bottom, sat a handful of black fleas. Every time they jumped they hit the thin metal lid. Ping. Pa-ping. Ping. Ping

Mystery solved.

As I set the jar down, my teacher walked up. "What're the fleas for?" I asked.

"I'm doing a little experiment," he replied.

"What kind of experiment?"

"Not sure. But if you want to find out, stop by the lab tomorrow and we'll see if it worked."

That night I couldn't sleep. The following day I skipped breakfast and ran all nine blocks to school. I tried to pay attention in my other classes but it was no use. We were on the verge of a major scientific discovery, and I, Rebecca Costa, devoted sixth-grade scientist, would be on the scene to witness the first incoming results! What could be more important?

When the bell rang, I raced to the empty lab where my teacher sat grading pig drawings. The jar was where I'd left it. "Have a seat," he pointed.

I sat down. The lab was quiet. No more pinging. The fleas must have died overnight.

Then he stood up and said, "Get down close when I pull the lid off." He set the jar on the edge of the desk, unscrewed the top, and crouched down so his face was even with the jar.

I got up and did the same. Eye-to-eye we watched the fleas jump up and down in the open container. Only this time—they jumped just shy of where the lid had been.

We stayed like this for several minutes—watching the fleas jump up and down until I couldn't stand it anymore. "Why don't they just jump out???"

He thought for a moment.

"Well... it hurts to keep banging your head over and over. So after a while, you stop doing it. But sometimes that's a trap. The fleas are free. They just don't know it yet."

In 2011, the popular American quiz show *Jeopardy!* laid down the gauntlet. After much back and forth, they invited IBM to pit their smartest computer against Jeopardy's most successful contestants: seventy-four-time winner Ken Jennings and the highest prize winner in the game show's history, Brad Rutter. It was to be a live showdown on prime-time television and IBM executives leapt at the opportunity.

But the scientists back at the IBM lab were not quite as optimistic—causing them to wonder if the executives had ever watched a full episode of *Jeopardy!* The game show wasn't as straightforward as they proposed...

For starters, the categories and questions were random, spanning everything from poetry and politics to perennials and postal codes.

And then there was the reverse format the program used: *Jeopardy!* provides the contestants the *answer* to a question, for which the contestant must formulate the correct *question*. So, for example, if "The capital of the Ottoman Empire" appeared on the board, the correct contestant response would be "What is Constantinople?" Likewise, the correct response for "A conceited fellow from Copenhagen" would be "Who is a vain Dane?"

Not an easy feat for a machine.

In the days that followed it became clear a computer would have to interpret the meaning of every word—used in a wide variety of contexts—then reverse-engineer its understanding of those words into something that could be "searched," then vet billions of potential responses, then rank those responses from the least to most probable, then select the most probable response and convert it into a question (*Where is a vain Dane? When is a vain Dane? What is a vain Dane?*).

Oh, and did I mention do all this before the other contestants rang in?

As the date for the prizefight grew near, the programmers grew anxious. The correct answers often involved a pun, a rhyme, or popular fad. Sometimes a metaphor, a quotation, folklore, passages in literature, and lyrics to a pop song appeared on the game board. Other times food recipes, sports trivia, lines from famous movies, and dates in history were the subjects. How could a computer make sense out of such a mish-mash of information? Especially a computer in an embryonic stage—one which, according to one developer assigned to the project, "could barely beat a five-year-old… let alone a Grand Champion."

Enter Dr. David Ferrucci, Senior Manager of IBM's Semantic Analysis and Integration Department, and the humble beginnings of Watson and Big Data technology. Ferrucci challenged a team of twenty-five researchers to discover a way in which natural, everyday language could be used to perform an intelligent Internet search and rapid analysis, then quickly pinpoint specific information and solutions.

By the time the much-anticipated televised match of man against machine rolled around, Watson had digested over two hundred million pages of information occupying 4 terabytes of storage space. Data from dictionaries, newswires, encyclopedias, taxonomies, and a vast cross section of sources were used. But this was a miniscule sampling of the information Watson needed to win. Studies reveal the human brain has a storage capacity of around 2.5 petabytes—roughly the equivalent of storing three million hours of television programming. Which meant that, together, Watson's human competitors would be playing with upwards of *five quadrillion bytes* of information against Watson's *four trillion.*

The human contestants had a five thousand to four advantage.

Given the size of the handicap, the IBM developers were hopeful Watson would have access to the Internet during game play. An idea Jeopardy quickly shut down. The players weren't allowed to have Internet access, so why should Watson? From their standpoint, it wouldn't be fair.

Fair?

In spite of the disadvantage, during the first couple of trials at the IBM lab, Watson faithfully rang in against mock players, isolated the most likely answers (reformulated as questions), and selected the response that had the highest likelihood of being correct. It did this routinely, without fanfare. Even when the developers threw in trick questions, somehow Watson found a thread and gracefully navigated around the deception. By all accounts it looked like Watson was ready.

Game on.

On February 14, 2011, the first historic match featuring Watson and the top winners of *Jeopardy!* aired on national television. It was an up-and-down, nail-biting match—scattered with a few humorous moments when Watson misunderstood the question and offered a bizarre answer. After round one, Watson tied Rutter and beat Jennings. By day three, Watson emerged victorious, easily laying claim to the $1 million prize—a prize IBM promptly donated to charity.

Meanwhile, back at the shop, the Watson team realized they had wandered into something bigger than a computer that could win a game show—they'd discovered a way *to convert the sum of human knowledge into simple, multiple-choice options.*

In the same way our earliest ancestors captured, corralled, and tamed wild animals for our collective advantage, the inventors of Watson threw a leash around the Information Age. And with that, foreknowledge and preemption took a quantum leap.

The lid was off.

Big Leap for Big Data

By the beginning of the twenty-first century, Big Data and predictive analytics had found its way into big information technology (IT) organizations, academia, and governments. For decades, those in charge of IT had been grappling with controlling, tracking, storing, distributing, analyzing, and protecting ever larger sums of data—never mind making it easy to get to the right information at a moment's notice or piecing together the data to gather useful insights. The problem was plain and simple: the Information Age had overshot its goal. More data was being generated by more people, on more devices, in more formats than anyone could get their arms around. We were no longer searching for a needle in a haystack—we had graduated to searching for a specific needle among *stacks of needles*—a problem technologists soon began describing as the Four V's: the *velocity*, *volume*, *variety*, and *veracity* of information.

The first two V's—the *velocity* and *volume* of information— are easy to understand: today we produce more information, at a faster rate, than anyone can use. Think of everything we know about the decisions we must make as being on a giant ticker tape—a tape relentlessly tick, tick, ticking away every nanosecond of every day. Changes in energy prices? Tick, tick, tick. New regulations? Tick, tick, tick. Announcements by competitors? Tick tick, tick. Medical breakthroughs? Tick, tick, tick. Terrorist threats? Tick, tick, tick.

Eric Schmidt, the Executive Chairman of Google, explains the challenge volume and velocity presents:

> There were 5 exabytes of information created by the entire world between the dawn of civilization and 2003. Now that same amount is created every two days.

When the executive chairman of one of the largest search engine and email companies in the world says we're overloaded, we're overloaded. Put succinctly, there is no other way to *metabolize* information but to leverage "smart" technologies to get to the essential.

Then there's the third V—the *variety* of information we produce. Did you know that every minute of the day, two to three days' worth of new video content is uploaded to YouTube? And while that sounds like a lot of video footage, it pales in comparison to the number of text and email messages we produce. Internet statistics site Pingdom claims we now send 144 billion emails a day. And in 2012, Harvard University reported we post a billion social media messages every couple of days.

Now add to this encrypted communications, retail transactions, telephone conversations, radio and satellite communications, etc., and it's easy to see why analysts estimate that over 95 percent of the information we create is "unstructured." Getting our arms around all this unstructured data has proven to be almost as challenging as designing a computer that could win *Jeopardy!*

And lastly, there's the *veracity* of data.

These days, for every study we find on the Internet that says one thing, there's another stating the exact opposite. And who has time to dive into the research methodologies or original source material used? Or corroborate by finding second, third, and other independent sources?

Against this backdrop, the IBM team got busy. Watson's performance on *Jeopardy!* may have marked a computing milestone, but the rigorous work of making a commercially viable machine that could tame the Four V's lay ahead.

Watson's first stop was health care. And they had good reason to start there: recent studies showed that physicians would have to spend 160 hours a week reading medical journals to remain current in their field. Which—when you do the math—leaves the sum total of one hour per day for sleeping, eating, family, *and* treating patients. Ridiculous!

But there's more...

That same study showed that 81 percent of physicians admitted spending five hours or less per month reviewing research. Which means doctors are falling behind every hour of every day. Dr. Steven Shapiro, Chief Medical and Scientific Officer at University of Pittsburgh Medical Center, shed light on the enormity of the discrepancy when he admitted, "Medicine has become too complex [and only] about 20 percent of the knowledge clinicians use today is evidence-based."

Twenty percent based on *evidence*?

Assuming he's right, what's the rest based on? If ever there was a need for Watson it was in health care—so Big Blue jumped in headfirst.

They began with the Memorial Sloan Kettering Cancer Center and WellPoint's oncology group. By accessing everything that was known about specific cancers, Watson proved it could quickly produce a prioritized list of probable patient diagnoses along with the most successful treatment protocols—the same way it prioritized correct answers on *Jeopardy!* Then, once the system listed the most likely diagnosis and best treatment option, Watson pointed to the next piece of patient data needed to enable the system to improve the accuracy of its analysis. This made it possible for a healthcare worker with little experience to know exactly what patient data to obtain next—then enter that information into Watson to see whether this altered the computer's findings.

In no time at all, doctors, nurses, intake personnel, and other hospital workers found themselves having a live *exchange* with the most knowledgeable oncology expert in the world—the only expert who could keep up with 160 hours of new medical information every week and never needed a break.

According to Dr. Samuel Nessbaum of WellPoint, Big Data has revolutionized oncology. Watson's hit rate for diagnosing lung cancer is now better than 90 percent. To put that into proper perspective, the rate for diagnosing lung cancer by physicians is around 50 percent.

So, now let me ask you this... after hearing that statistic, *who do you want to diagnose you? Suggest a treatment plan for you?*

Today, if you're fortunate to be admitted to a hospital that has a Watson-like Big Data system, your survival is no longer dependent on which doctor you happen to be assigned to. Or whether that doctor got enough sleep the night before, had time to eat dinner, just argued with their spouse, or had a fender bender on their way into work. Big Data systems now bring *all medical knowledge* to bear on a single patient, at a single moment in time, without any of the disadvantages of their human counterparts. In much the same way CAD systems elevated every engineer's ability to design safer, better products by averting failure, even the most adept physicians are made more potent by fast analytics.

Two years after the historic challenge on *Jeopardy!*, it was difficult to find an industry that was not making the move to Big Data: banking, meteorological, pharmaceutical, energy, and agricultural institutions all jumped on board. By 2013, revenues from Big Data jumped 58 percent, climbing to $19 billion, and the market is on course to eclipse $150 billion in 2017. All of which means it won't be long before a doctor in a remote village in Africa will have the same computer-assisted decision-making abilities as a trader on Wall Street. And will be able move just as quickly to circumvent danger.

Predictive Analytics Comes of Age

Big Data systems not only tore down data silos that prevented us from using all available information to our benefit, but also launched a new era of predictive modeling. And with the capability to identify new patterns, our foresight grew more consequential.

Then in 2009, Swedish-American start-up Recorded Future set out to achieve the impossible. A small team of scientists began working on algorithms which searched publicly available information for obscure correlations, relationships, and patterns that might foretell an unexpected event. But not just any event. A *major* event, such as a government coup, financial collapse, or terrorist attack. Was Big Data and predictive analytics up to the task?

The company began with four simple assumptions:

First, that past and present behaviors were indicators of future behavior.

Second, the amount of public information was growing so rapidly, there was no need to invade anyone's privacy.

Third, the technology to search, aggregate, and analyze the data in the way they needed it to existed.

And lastly, humans broadcast their intentions prior to acting out—a habit professional poker players have become so familiar with they've invented a word for it. They call it a "tell."

With this in mind, Recorded Future went on a hunt for "cyber tells."

The company moved quickly. Within a year, their algorithms began finding astounding relationships between information on social media, emails, blogs, videos, photographs, and 650,000 other public sources—connections no one had previously looked at as a whole. But that was just the tip of the iceberg. The company then laid that information across a proprietary past, present, and future timeline to see if that timeline could foretell what would occur next, and when.

And in 2010, Recorded Future hit pay dirt.

They predicted Yemen was headed for upheaval.

Co-founder Christopher Ahlberg was elated. "Yemen took five months longer than we predicted—but if you go back and look at our earliest blog posts—it's ALL there." He continued, "We're trying to find new ways of generating data that tells us what is going on in the world—what did happen—and *what will happen.* We're not going to get 100 percent in terms of outcomes, but we can pull things together in a way that no one else can."

Today, Recorded Future supplies analysis to hundreds of exclusive subscribers, which include hedge fund managers, state and defense departments, and the CIA. They sound the alarm when the data foretells economic turbulence, cyber threats, or a forthcoming coup—offering the single most powerful advantage to their clients: the opportunity to get a jump on the future.

Predictive Means Proactive

It's impossible to talk about man's conquest of the future without talking about Fuzzy Logix. As news spread that 60 percent of drug overdoses in the United States were attributable to prescription opioids, and healthcare costs associated with opioid addiction reached $25 billion, Fuzzy Logix announced a predictive model that could determine whether a patient was likely to abuse opioids before they were handed their first prescription. Using a patient's "past medical and pharmacy utilization, location, and demographic data," Fuzzy Logix can identify patterns that—according to Blue Cross Blue Shield of Tennessee—permit physicians to eliminate candidates for opioid painkillers 85 percent of the time.

And poof. Just like that. Eighty-five percent of the problem gone. But there's more.

Today, Fuzzy Logix offers more than seven hundred predictive algorithms, including the most accurate analysis for predicting type 2 diabetes, as well as sufferers who are not likely to follow treatment protocols for the disease. By analyzing telltale markers, these algorithms arm doctors with the ability act proactively—to better design treatments to fit each patient's predispositions. And algorithms that identify the risk for heart attacks, strokes, and other diseases are also on the way. But just in case the impact predictive analytics is having on health care isn't enough to get you excited, Fuzzy Logix offers models for credit risk management, supply chain optimization—even marketplace resistance to a new product or service. The company's ability to find data patterns, then use these patterns to identify outcomes in a broad range of disciplines is unsurpassed. And every day their accuracy grows—until one day there will be an algorithm for every future likelihood.

In the meantime, on the other end of the country, Gene Tange, founder of PearlHPS, had a similar idea. What good is all the data we're generating if it can't be used to avoid failure? After twelve thousand hours of laboriously studying high-performing companies, Tange uncovered the reason 70 percent of new business initiatives were doomed before they ever got out of the starting block—and of those that did make it, 56 percent failed to deliver the value expected. The problem was *execution*. Businesses might have had a strategy, a plan, and the resources in place, but that was just a starting point. There were other critical variables. So Tange and his team began outlining the parameters that separated the successful introduction of new products, procedures, technologies, etc., from those that bottle-rocket. And, following several lean years—and self-doubt as to whether such a formula was even possible—PearlHPS emerged with Execution Analytics, a working algorithm that could predict business outcomes twelve months in advance—an algorithm Steve Cadigan, former VP of Talent at LinkedIn, called "the holy grail of analytics."

Using PearlHPS' Execution Analytics to compare and contrast the consequences of their options, one company cut their new product introduction time by nearly 20 percent, generating $150 million in early-release revenues. In another example, a tech company torn between continuing down the path they were on, partnering with a larger firm, or selling their company entirely was able to assess the financial and developmental risks associated with each alternative in a way the founders and executive team could not.

As companies like Recorded Future, Fuzzy Logix, PearlHPS, and others leverage Big Data to add more variables to their algorithms, we are beginning to realize the full implications of predictive analytics. We know which twelve-year-olds are likely to become binge drinkers, which employees are prone to cheating, and which prisoners will reoffend. We know when civil unrest is brewing, when rivers will flood, and how fast a virus will spread. We see these things with a clarity previously reserved for gods.

But we are not gods. From institutional resistance to fear and genetic imperatives, foreknowledge faces powerful adversaries.

"It is no good to try to stop knowledge from going forward. Ignorance is never better than knowledge."

- Enrico Fermi

CHAPTER FOUR

Unintended, But Not Unanticipated

By the time my nephew, Camden, turned two we knew he was not like the other boys. He had food issues.

Left unchecked, he would pry open the pantry door and climb the shelves to get at his favorite snacks. It was also common to round the corner to find him eating ice cream with a spoon straight from the freezer drawer—or circling a pie on the counter, breaking off pieces of crust and shoving them into his mouth as he walked by.

It didn't take long before every cabinet in the kitchen was on lockdown, including a deadbolt my brother installed at the top of the pantry door. Which worked fine, until one day he spotted Camden shoving a chair toward the deadbolt.

"We have a problem," my brother sighed.

"You mean with food?"

"Yeah. We can't go anywhere without Camden begging for candy and soda. Everywhere we go he's grabbing candy bars off the shelves. It's gotten really bad."

"Maybe you shouldn't take him with you... you know... for a while..."

"That's not the worst part. We started toilet training last month. At first we put his name on the refrigerator and we let him put a star next to it when he let us know he had to go. But he didn't care about the star. So then we got this idea to give him an M&M when he let us know."

"An M&M?"

"Yeah, just one."

"Did it work?"

"Like a charm. In two days the training pants were off. He'd run and get us right away. And when he finished we gave him his M&M."

"So he's out of diapers—congratulations!"

"That's what we thought. We were both patting ourselves on the back. Except then we started noticing that Camden was going to the bathroom, and then a minute later, he had to go again."

"What do you mean?"

"He was letting just enough out to get an M&M. Then he'd wait a couple of minutes and come get us so he could get another one. Now we're running back and forth to the bathroom and we're out of candy."

"He's metering it out?"

"Yep. He's only two and he's working the system..."

If we thought the ability to manipulate outcomes was all upside, then we don't know human nature. The same tools that allowed engineers to identify problems before the fact paved the way for a different kind of success...

Once we had the ability to analyze millions of potential scenarios ahead of time, the responsibility for prevention fell squarely on the shoulders of businesses everywhere. And, true to form, a mountain of liability lawsuits ensued. All a plaintiff had to show was that a company *could have foreseen* danger—the product itself need no longer be faulty.

Claire Andre and Manuel Velasquez, professors in Business Administration at Santa Clara University, were among the first to sound the alarm:

> According to strict liability laws, a manufacturer can be held liable for injuries even when he or she had no way of preventing those injuries. Holding manufacturers responsible for injuries caused by products known to be defective or potentially dangerous is one thing, but today manufacturers face lawsuits—often bordering on the outrageous—*for injuries they could not have prevented.*

"Could not have prevented?" Can a fast food chain *prevent* me from spilling coffee on myself? Can a gas barbeque maker *prevent* me from using it as an indoor heater? Can a company that manufactures children's car seats *prevent* me from incorrectly installing the seat?

According to the law they can.

The fact is, as computers grew better at spotting problems, the more vulnerable businesses became. With so much data at a company's fingertips—with so many ways to safeguard against every conceivable danger—it became easy for plaintiffs to prove how a business coulda shoulda acted. And to the untrained juror, coulda shoulda sounded an awful lot like *negligence*.

One internal email from an employee suggesting a safety precaution; one expert demonstrating how the technology to foresee the problem existed; one watchdog group claiming stronger warnings might have prevented misfortune; one victim with burns, missing limbs, or worse yet, the loss of a loved one was all it took for plaintiffs to prevail, and the number of liability suits to boom.

LexisNexis Market Intelligence reported that in 1999, nearly five thousand personal injury product liability cases were filed in US Federal Court. Within five years that number doubled. Two years after that, caseloads doubled again, reaching twenty-eight thousand claims, causing Gary J. Spahn, then co-chair of the American Bar Association's Litigation Section Product Liability Young Lawyers' Subcommittee to say, "I don't see a letup anytime soon."

But it's not just the number of suits. Settlements also shot through the ceiling. In a study by Jury Verdict Research, the median award for product liability cases jumped from $550,000 in 1985 to over $1.5 million in 2011. By the time 2014 rolled around, the average award had climbed to over $5 million. To put that figure into perspective, the average medical malpractice jury award during that same period was less than half the amount paid for product liability cases.

Medical malpractice!

But businesses were not about to stand by and watch their profits decimated by a tsunami of coulda shoulda lawsuits. They countered by doubling down on liability insurance. In turn, insurance companies loaded up with high-powered attorneys who were determined to fend off class-action suits and inflated penalties. This produced expensive, protracted litigation, which in turn triggered an unstoppable rise in premiums. As the spiral of nonsensical claims, costly suits, lucrative settlements, and high premiums gained steam, these costs were eventually reflected in the price to the consumer. Andre and Velasquez sum up the damage:

> About 60 percent of the average corporation's litigation expenses today are product liability cases. In a recent report by the Conference Board, 15 percent of corporations surveyed had laid off workers because of product liability costs, while 8 percent had been forced to close plants altogether.

Blitz USA was one of the fatalities. Following a successful forty-six-year run, the largest manufacturer of red, portable gasoline containers filed for bankruptcy in 2012. The reason? A handful of consumers thought pouring gas on an open fire was a good idea. The gas vapors ignited and the containers exploded.

The liability suits claimed Blitz should have *foreseen* misuse of their cans. The company should have implemented shields at the mouth of the containers to prevent "flashbacks." But, according to Blitz CEO Rocky Flick, shields would not have prevented the explosions that occurred. "There was no proven device that we could get that we thought would prevent somebody from getting hurt when they elected to pour gasoline on a fire."

But Flick missed the point.

Attorneys for the plaintiffs didn't care whether the shields worked. And they didn't care whether Blitz included warnings that gasoline was flammable and should never be placed near fire or an accelerant. Or that even a third grader knows better than to put a match to gasoline. Blitz had an obligation to *try* to prevent individuals from pouring gas over fires. It was the lack of effort by Blitz the attorneys went after: Blitz's decision to put profit ahead of making product changes—any changes—even if those changes would not have protected a single victim.

Emory University School of Law professor Frank Vandall takes a pragmatic view of Blitz and other companies facing similar exposure: "There is no way you can avoid liability for a can like this, because *there is going to be injury*, and when there is injury, there are going to be lawsuits." Which begs the question—if the plaintiffs had used a bowl to pour gas on the fire, would they sue the bowl maker for not acting preventatively? How about a watering can? A cup? If the product itself isn't faulty, where do we draw the line?

In Blitz's case, the line was unclear. The company sold more than fourteen million containers, and experienced fewer than two problems for every million sold. Yet, in spite of a safety record that puts the Samsung Galaxy Note 7's exploding battery to shame, the cost to defend Blitz was far and above what the company could bear: $30 million in legal fees, an additional $30 plus million in insurance company settlements, skyrocketing premiums, lost time and wages of management, and more.

It turns out foreknowledge comes with a price. And that price was bigger than we knew.

One of the most egregious, best known liability cases occurred in 1994. The story goes that senior citizen Stella

Liebeck ordered a forty-nine-cent cup of coffee from a McDonald's drive-through window. Liebeck was sitting in the passenger seat and placed the coffee between her knees to remove the lid to add sugar. While removing the lid, the coffee spilled on her cotton pants, causing third-degree burns.

After a lengthy trial, the jury determined that McDonald's was 80 percent responsible. The company failed to warn customers the coffee was hot. According to the jury, McDonald's had a responsibility to *foresee* the danger associated with hot beverages. They awarded Liebeck $200,000 in compensatory damages and an additional $2.7 million in punitive damages.

A later court reduced this settlement—but only after Leibeck's windfall opened a floodgate of similar suits.

In 2009, a suit was brought against the manufacturers of Bluetooth headsets for failing to warn consumers that turning up the volume could cause hearing damage. Really? Are there people who don't know loud sounds are harmful? Apparently there are.

Enough for twenty-six suits to be filed.

Then there is the story of twenty-seven-year-old Daniel Dukes, who, in 1999, deliberately hid inside Orlando's Sea World until it closed. When everyone left and the park was quiet, he dived into a tank to swim with a killer whale and died. Duke's parents sued Sea World on the grounds that they did not display warnings that a killer whale could kill.

Evidently the name "*killer* whale" was insufficient.

In a humorous article, journalists Brett Nelson and Katy Finneran demonstrate just how difficult getting our arms around every possibility has become. They note that warning labels have not only expanded in size but now border on absurd:

- Rowenta iron: "Never iron clothes on the body."
- New Holland tractor: "Avoid death."
- Nytol sleeping pills: "May cause drowsiness."
- Peanut M&Ms: "This product may contain nuts."
- Vidal Sassoon hairdryer: "Do not use while sleeping."
- Razor scooter: "This product moves when used."
- Apple iPod Shuffle: "Do not eat."
- Zantac 75: "Do not take if allergic to Zantac."
- Hormel pepperoni: "Do not eat packet."
- Marks & Spencer bread pudding: "Product will be hot after heating."

Yesterday I noticed my hairdryer came with a warning not to use it when I am in the shower. A white tag hanging from the cord showed an illustration of a person standing under a running showerhead using their dryer. There was a large red X over the drawing. Unless you're of the opinion that people have completely lost their minds, these labels aren't only humorous—they're insulting.

The more our ability to avert danger grows, the greater the burden on prevention. Smokers sue tobacco companies. Diabetics sue junk food makers. Homeowners sue for mortgage contracts they didn't read. What happened to personal accountability? Is it reasonable to expect manufacturers to control bad decisions and poor behavior? More to the point—what do we do when the data we have is vague or *inconclusive*? When there's enough information to suspect a problem could exist, but not yet enough to act on?

What then?

Junior Seau's Brain

As is often the case, we may have the foresight to see a problem brewing, but not the precision, knowledge, or cure to nip it in the bud. Often a solution has such large implications, it makes it difficult to argue on the side of caution. Other times there is conflicting evidence and opinion—muddying the waters to the point where we have no alternative but to wait until there is additional data one way or another. Institutional resistance, politics, and self-interest are also obstacles that prevent us from acting.

Take the case of professional sports and chronic traumatic encephalopathy (CTE), for example.

In 2012, on an unseasonably warm spring day, the body of American football star Junior Seau was discovered in his San Diego, California, home. Seau, just forty-three years old, died of a self-inflicted gunshot wound to the chest, adding to a growing number of National Football League (NFL) players who have committed suicide.

In 2006, defensive lineman Shane Dronett began complaining of episodes of confusion, rage, and paranoia. According to friends and family, he battled the condition for three years prior to ending his life. Pro-Bowlers Dave Duerson and Andre Waters met similar ends. As did young Penn State lineman Owen Thomas.

While suicide is seldom traceable to a single cause, these athletes all shared one thing in common. They suffered from a neurodegenerative disease called chronic traumatic encephalopathy (CTE).

CTE occurs when a protein called *tau* builds in the brain after repeated blows to the head. The effect CTE has on behavior has been documented: confusion, depression, aggression, memory loss, problems with impulse control, and impaired judgment. But it wasn't until we acquired modern imaging technology that we could observe the progressive damage that results from multiple concussions: forty-year-old athletes who have CTE appear fit on the outside, but their brains resemble ninety-year-olds. What's more, the condition is not reserved for football players alone. Athletes who play soccer, rugby, and hockey also reveal disproportionate evidence of CTE.

To date, more than five thousand players have sued the NFL over CTE. Recently, seventy-five athletes brought a lawsuit against helmet maker Riddell Sports, Inc., claiming both the NFL and Riddell concealed medical information about the long-term dangers of concussions from players, coaches, and trainers for eighty years.

But when it comes to CTE and sports, not all of the data is in. It's more accurate to say that we are in that messy, in-between period where we have enough knowledge to see there is a relationship between head trauma and CTE, but not enough to know *which* blows present a danger. Or which athletes will be affected. Or when they will be affected. After all, not every person who gets a concussion develops CTE. And, not every head blow produces CTE. And not every athlete who suffers from CTE shows symptoms or commits suicide.

This makes preemption tricky.

Even trickier still is the fact that the only way to definitively diagnose CTE is by conducting a postmortem. In other words, a player has to die first. How do we respond to a disease that cannot be diagnosed while a person is alive?

Another challenge that makes it hard to act on CTE is that we have no way to know the g-force of any particular encounter on the field. And without that key piece of information, there's no way to know how hard a player was struck. Or whether the strike had any short *or* long term effect on the brain. Since we can't look under the cranium of every player after every collision, the best a team physician can do is *guess.*

And that is what they do. They ask injured players a few questions, check things like balance, recollection, speech, comprehension, pupil dilation, etc., then make their best call. If they're smart, the physician errs on the side of caution—even if the score is close and there's pressure to keep the athlete in the game. Even if the player acknowledges the danger and insists on continuing. Even if the physician earns a reputation for being too cautious and finds themselves out of a job.

Even if.

And finally, there's the game itself. Players are being hit every second of every minute. Similar to boxing, head blows are part of the sport. So much so that the Boston University School of Medicine CTE Center reported that the early signs of CTE have been spotted in fourteen out of fifteen former NFL players studied. If players were removed from the field and given a full physical every time they got hit, fans would be headed home after the kickoff.

But now that an alarming connection between football, concussions, depression, and suicide has been discovered, what do we do? Now that we can visibly see and chart the devastating long-term consequences of CTE, what responsibility do team owners, the NFL, equipment manufacturers, and the public have? Should we assume athletes understand the medical risks they're exposed to when they sign medical waivers and let that be that? Should the sport be ended on the grounds it is deadly? Or, similar to manufacturers besieged with lawsuits, should sports franchises and athletes simply double down on insurance and bury the costs in ticket sales and bigger contracts? Here, journalist Jonathan Tamari cuts to the chase:

> What do we do with that information?...Does it change whether parents let their children play football? Does it change how we react when a player such as DeSean Jackson returns from a concussion? Will the focus be on what long-term damage he may have suffered, or whether he'll go strong over the middle?

He continues:

> Players make their own decisions. This isn't about their decisions. This is about those of us who choose to watch and cheer and write. Do we act unaffected, even as the consequences become more and more obvious? Or do some fans turn away? Could it be enough that it dents football's standing as the king of American sports?

Not likely. The NFL is a $75 billion franchise. It isn't going to shut its doors anytime soon. What's more believable is the scientific evidence linking head trauma to CTE and CTE to sports will grow stronger with time, and, one by one, the lawsuits will be settled. It's already started. In 2015 a federal judge approved a $1 billion settlement by the NFL that allots $5 million per retired player for "medical conditions associated with repeated head trauma." And though the NFL's official position had previously been "There were simply not enough studies or medical proof for the league to make a direct connection between football and CTE," in 2016, testifying before the US House Committee on Energy and Commerce, NFL Senior Vice President of Health and Safety Policy Jeff Miller admitted there was sufficient data to establish a "link between football and degenerative brain disorders like CTE."

Unfortunately, neither a settlement nor admission by the NFL will diminish the number of athletes who pay the ultimate price for victory. This is what's frustrating about foreknowledge. Our view is limited until we get to the top of the hill. So our ability to see the big picture depends on how fast we climb. To that end, we are climbing faster than ever before. For example, there's no question that breakthroughs in sensor technology and motion analytics will soon allow physicians to pinpoint the precise g-force of every hit during gameplay. This, along with portable diagnostic equipment on the field, advanced genetic profiling to determine a player's inherited susceptibility for a spectrum of cognitive diseases, screening for chemical pre-CTE markers, and other technologies, will all help to reduce the number of cases. In time, scientists will garner a better understanding of the tau protein responsible for CTE and how to guard against it with new pharmaceuticals and other prophylactic treatments. Safety equipment makers will glean new insights from in-depth neuroscientific studies and state-of-the-art composite material research, as well as from other industries concerned with head trauma, permitting them to introduce the next generation of head and body armor.

It's not perfect. But then this is how progress unfolds. It happens in fits and starts. Sometimes we have to climb down the hill we're standing on and start at the bottom of a higher one to get a better view.

Trouble with Terrorism

If getting our arms around CTE has been abstruse, stopping terrorists *before the fact* has put every law, institution, and value free societies hold sacred to the test. Even with the most sophisticated surveillance technology and analytics at our disposal, we often find ourselves connecting the dots after a terrorist strike—causing many to ask "Why didn't authorities do something beforehand?" After all, modern terrorists are public with their opinions and intentions—often posting threats on social media and YouTube videos and committing smaller, telltale crimes along the way. And every attack requires advanced planning and logistics, which means supporting actors, communication, and coordination. We now know there *are* advanced signs—many clues that point to an attack in the making.

But are they enough?

Not in the case of Abdelhamid Abaaoud.

It was a typical Friday evening in November 2015. People in every town and city were looking forward to wrapping up the work week—stopping off for a bite to eat, gearing up for a ball game, making plans to listen to live music, or stopping by the market to load up on supplies for the weekend. Early birds were starting their holiday shopping, restaurants were turning tables as fast as they could, and taxi drivers were settling in for a long night. The streets were bustling in the Saint-Denis suburb of Paris.

Then at 9:20 p.m., tragedy struck. Terrorists opened fire at six popular gathering places, killing 130 and injuring 368.

It was the worst attack on French soil since World War II. President François Hollande declared a state of emergency, calling the senseless violence "an act of war."

French and EU authorities wasted no time working backward to identify seven of the eight perpetrators who, by this time, had blown themselves up or been shot by the police. But the mastermind, Abdelhamid Abaaoud, slipped by them. And there was reason to believe he was headed for Belgium's capital, Brussels. Learning this, Belgium's leaders immediately shut down schools, government offices, and businesses, warning citizens to stay off the streets. Army convoys were dispatched to buttress local law enforcement, and for three days and nights the city went into lockdown.

Using DNA, fingerprints, passports, rental car information, apartment leases, job references, and thousands of other clues, forty-eight hours after the attack, the French government had the names, backgrounds, and recent activities of the terrorists and supporting actors. The police raided two hundred locations, arresting twenty-three radicals and placing 104 others under house arrest. Accomplices were also rounded up in Belgium, Germany, Greece, the United Kingdom, and elsewhere. No one was taking chances.

By the following Monday, President Hollande felt satisfied the danger was contained and lifted the state of emergency. Parisians returned to work, shops and restaurants reopened, and the city did its best to put the nightmare behind them.

But for the French General Directorate for External Security (DSGE), the work was just beginning. Why didn't they see this coming? What had they missed?

As the investigation unfolded, it became clear the attackers had foreshadowed their intentions long before the first shot was fired. For example, authorities already had a lengthy dossier on Ismaël Omar Mostefaï, the twenty-nine-year-old French suicide bomber who murdered eighty-nine people at the Bataclan theater. Mostefaï had been identified as a dangerous Islamic militant and had eight previous arrests to his name. Eight.

The two attackers responsible for setting off bombs at the Stade de France sports arena had been living in Belgium where authorities had also been surveilling them.

And when it came to the leader of the Paris massacre, the early warning signs were all there, right out in the open.

Abdelhamid Abaaoud was the child of Moroccan immigrants who moved to Belgium to give their children a better life than they could in their homeland. He grew up in the multicultural neighborhood of Molenbeeck-Saint-Jean and was fortunate to attend the prestigious school of Collège Saint-Pierre d'Uccle. By all accounts Abaaoud was a good student, a helpful son, and had what most would consider a normal childhood.

But sometime during his impressionable teenage years, Abaaoud was recruited by Belgium's most notorious Islamic extremist group. Young and eager to make his mark, he took an oath to become a soldier in a holy war—a war he initially knew little about. Abaaoud made no secret about his affiliation. In a chilling 2014 video he posted on the Internet (for all to see) he claimed, "All my life, I have seen the blood of Muslims flow, I pray that Allah will break the backs of those who oppose him... that he will *exterminate* them." In another video Abaaoud shot in Syria, he is seen tossing bloodied corpses into a trailer, saying, "Before we towed jet skis, motorcycles, quad bikes, big trailers filled with gifts for vacation in Morocco. Now, thank God, following God's path, we're towing apostates—infidels who are fighting us."

As alarming as these and other videos are, they fall short of the evidence required to detain or bring charges—not in a society that protects free speech. Though it was apparent he was getting bolder, more agitated, more dangerous, there was little law enforcement could do but continue to watch...

Then, much later, we discovered that Abaaoud was responsible for several failed attempts before Paris. Ten months prior to the massacre, Belgium authorities broke up a terror cell in Verviers that Abaaoud had financed and organized. Though his picture was plastered all over television, the Internet, and newspapers, Abaaoud bragged to one reporter, "I was even stopped by an officer—who contemplated me so as to compare me to the picture. But he let me go." He escaped the Belgium raid and made his way back to Syria.

But there was more.

Six months before Paris, Abaaoud was linked to the Paris-bound train where a member of ISIS opened fire on passengers—an assault thwarted by three American passengers who overpowered the gunman. Authorities also tied Abaaoud to the attempted attack on a church in the Paris suburb of Villejuif as well as several other thwarted efforts.

In the days following the tragedy in Paris, the evidence against Abaaoud piled up, leaving little doubt that he had been climbing the leadership ranks of ISIS, growing ever more resolute. Secret communications over social media, associations with other terrorist cells in the United Kingdom, more threatening videos, and a multilayered trail of money and arms were discovered. Most of this information was easy to find—that is once investigators knew what to look for. Regrettably, it took the lives of 130 innocent victims to connect the dots in the right way. But for the friends and families of the victims, and those who survived the ordeal, it was far too little, too late.

Just about the time I was wondering why the police didn't piece together this evidence earlier, and whether that would have prevented the bloodshed, news broke that *Iraqi officials had given France warning that an attack was imminent twenty-four hours before it occurred.*

Iraqi leaders produced documents that showed the French government received and ignored an intelligence warning that ISIS leader Abu Bakr al-Baghdadi had ordered sleeper cells to launch attacks on coalition countries fighting in Iraq and Syria. The Iraqi warning contained specific information about the attack. It stated that the terrorists had received training in Raqqa, Syria, prior to entering France, and that a twenty-four-person cell in Paris was coordinating the assault: nineteen attackers were involved, along with five logistics personnel.

This was good intel. Actionable intel.

Then later, Turkish officials came forward, stating they had also issued warnings to the French authorities.

When confronted with the fact the French government had early knowledge, one high-ranking official admitted there wasn't much they could do with the information. "We get these kinds of warnings every day," he said.

And he had a point.

Getting hundreds of warnings every day is the same as getting none. If there is no credible way to vet them, how do you decide which to act on?

The second problem with the Iraqi warning is whether one piece of intel justifies trampling all over people's rights. Imagine asking for a search warrant for every person who had recently traveled from Syria to Europe and was living near or in Paris. By the end of 2015, nearly four hundred thousand Syrian refugees were on the move in Europe. Even if authorities were successful at convincing a judge to give them a "blanket" warrant—how many detectives would you need? How would you locate everyone who fit the profile described?

But Belgian and French authorities are not alone. Every country in the world faces the same quandary when it comes to preempting terrorism.

In 2004, ten bombs in backpacks were placed on four commuter trains in Madrid, killing 192 and wounding two thousand. An investigation revealed police informants had passed multiple tips of the attackers' plans to local authorities only weeks before the disaster. Equally heartbreaking was the fact that a small team assigned to monitor Islamic extremists in Spain were already performing surveillance on the perpetrators. The suspects were already known to be dangerous.

In 2011, thirty-two-year-old Anders Behring Breivik detonated a fertilizer bomb at a popular tourist resort off Lake Tyrifjorden in Norway, then opened fire, murdering seventy-seven vacationers. Following the mass slaying, police learned Breivik had been publishing fundamentalist Christian, anti-Muslim blogs on his website and making social media posts that showed the pot was boiling.

Likewise, the assassins responsible for the murders at the Boston Marathon, Virginia Tech, Sandy Hook Elementary School, Columbine, and Umpqua Community College previewed their intentions in advance on the Internet—some going so far as announcing their plans the day they followed through.

It's alarming to learn these executioners were hiding in plain sight—broadcasting their actions ahead of time on public forums, boasting without so much as covering their faces, acting out in pathological ways. Yet, from a pragmatic standpoint, government authorities cannot single out every person who writes a threatening blog or posts an angry video. They can't track every immigrant, radical Islamic worshipper, gun owner, and citizen who makes regular phone calls to the Middle East or buys fertilizer at the local hardware store. It's impossible for law enforcement to be in all places at all times.

Or is it?

Here again, technology holds the key to a safer future.

The repeated use of threatening words in typed and audio communications; the modality of voices; patterns in telephone activity, travel, credit card use, and purchases; associations and affiliations, and bank activities, can all be unobtrusively aggregated and analyzed, in real time, for early signs of trouble by today's Big Data systems—machines capable of stringing together billions of dynamically changing variables in a way, and at speeds, no team of humans can. By connecting the dots, these systems can identify behaviors that suggest a high probability of violence, then call this information to the attention of authorities *long before a threat comes to fruition.*

In the case of Abdelhamid Abaaoud, his involvement in previously thwarted attacks, his recent travel from Raqqa to France, his video and social media posts, the twenty-four-person cell he was affiliated with in Paris, his phone activities, his name on the list of dangerous suspects obtained by the Belgium mayor, and hundreds or thousands of other discrete pieces of data could have been mapped to reveal a dangerous pattern—one which, when combined with advanced Iraqi intel, could have quickly identified Abaaoud as a probable instigator.

Now add the power of Big Data analytics to new advancements in facial recognition.

Facial recognition?

Yes, facial recognition. Where evolution and the digital world meet...

When you think about it, human beings are designed to show our feelings, and foreshadow our intentions, through facial expressions. Though we're often unconscious of the fact that we're broadcasting our emotions, our terror, anxiety, sorrow, remorse, pleasure, and even our attempt to deceive are all right there for the public to see. Facial recognition software tracks the forty-three muscles in the face responsible for revealing these innermost thoughts—making it easier to spot individuals preparing to engage in nefarious acts. When used in combination with other clues such as darting eyes, the repetitive licking of the lips, rapid blinking, shifting the weight on our feet, unusual perspiration, and elevated body temperature—and patterns of behavior, activities, and movement unearthed by today's sophisticated Big Data systems—these indicators offer new opportunities to stop perpetrators before the fact.

Imagine cameras with facial recognition software pointed on every commuter, every shopper, every student, worshipper, laborer, or suspected terrorist. Now imagine alerting immigration officers or local police there is facial evidence of an imminent threat! The only question is: what will authorities do with this information? Will it rise to the burden of proof for crimes which have not yet happened? Will the law evolve to accommodate "preemptive law enforcement" based on computer analytics alone?

After all, the insights we acquire about the activities, behaviors, and expressions of terrorists won't add up to much if we don't have the legal authority to act on that knowledge. Foresight's value is measured by action. If we don't plan to act on what we can now decipher about future events, we might as well rely on a deck of Tarot cards.

Daycare Sociopaths

Our newfound foreknowledge has challenged the laws governing preemption in the same way that CTE has thrown a monkey wrench into professional sports. But these things pale in comparison to the controversy associated with knowing which youngsters are on the path to becoming adult sociopaths. Here again, we struggle with what we know and what can, and should, be acted upon.

In 1990, HBO released a disturbing documentary titled *Child of Rage*, the first film to chronicle children who exhibit early traits associated with adult sociopaths: the absence of empathy, manipulation, cruelty to animals, pathological lying, and so on. In the film, a parent describes her nine-year old son as "intelligent, cold, calculating, and explosive," while another mother tearfully says her son is "callous and unemotional," causing the rest of the family to fear for their lives. At one point in the film, a *six-year old* girl looks straight into the camera and explains the pleasure she derives from torturing and killing her neighbor's pets. She gleefully describes the pain she regularly inflicts on her siblings when no one is around—including sexually molesting them with objects. And she goes on to admit fantasizing about one day murdering her parents. Her matter-of-fact demeanor is chilling, leaving no doubt as to the seriousness of her intent.

One of the most alarming cases to be covered by the news involved nine-year-old Jeffrey Bailey, Jr.—an otherwise happy, cherubic-looking youngster with clear, bright eyes. On a sunny afternoon, for reasons he could not explain, Bailey pushed a three-year-old into a swimming pool, then pulled a lawn chair up to the edge of the pool to watch the toddler struggle and drown. Once drowned, Bailey calmly got up, returned the chair, and walked home to eat dinner. As far as Bailey was concerned, nothing unusual happened that day.

The story of Bailey and other child killers are not as rare as you might think. There are thousands of children who exhibit early danger signs.

- In 1984, Joshua Phillips' mother was cleaning his room when she discovered the dead body of their eight-year-old neighbor, Maddie Clifton, under his bed. Fourteen-year-old Joshua said he accidentally hit the girl in the eye with a baseball bat and panicked when she screamed. So he dragged her to his room and continued beating and stabbing her until she finally "shut up."

- In 1993, Jon Venables and Robert Thompson, both ten years old, took two-year-old James Bulger by the hand and led the toddler out of a shopping mall in Liverpool, England. Once away from the mall, they spent hours torturing him before beating him to death and hiding the body.

- In 2009, fifteen-year-old Alyssa Bustamante confessed to luring her nine-year-old neighbor Elizabeth Olten into a nearby forest and killing her. Bustamante wrote in her diary "...It was ahmazing. As soon as you get over the 'ohmygawd I can't do this' feeling, it's pretty enjoyable."

Setting aside for a moment the debate as to whether these acts are the result of nature or nurture, we can no longer deny the fact that we can now identify a sociopath in the making. The more information we garner from studying adult offenders such as Ted Bundy, John Wayne Gacy, Jeffrey Dahmer, and Edward Gein, the more we learn what to look for. And today, the symptoms leave little doubt.

Of course, the obvious problem with accurate diagnosis is, even if we know the symptoms, we don't yet have a cure for sociopaths. Similar to CTE—we have the foresight to see what lies ahead, but no real way to stop it.

What to do, what to do...

Dr. Paul Frick, a psychiatrist specializing in child psychopathy at the University of New Orleans, believes we must refrain from labeling children until we have a definitive, 100 percent fool-proof test. And maybe even then. From Frick's perspective, traits such as narcissism and a lack of impulse control are attributes also found in healthy children. Therefore, attributes associated with psychopathology is "a matter of degrees."

And we all know that as soon as anyone claims it's "a matter of degrees," a heated debate is guaranteed to ensue.

In her recent article "Can a Kid Be a Psychopath?" Lylah M. Alphonse describes the problem the psychiatric community now faces:

> Experts are divided about whether it's right to label a child as a psychopath. On the one hand, their brains are still developing; since psychopathy is largely considered *untreatable* such a label would carry a heavy, life-altering stigma. On the other hand, identifying "callous-unemotional" children early would allow for successful treatment—or at least a heads-up to society.

On the one hand, early diagnosis carries a life-time sentence. On the other, do we wait until there's a dead body? How different is this from the knowledge that a blow to the head can lead to suicide? Or evidence that a terrorist is plotting an attack? In each case we can foresee danger with much greater accuracy than ever before. In each case there is a pattern which leads to an inevitable conclusion. In each case we have the opportunity to stop a moving train. That said, a high probability is not the same as certainty. Though we are rapidly headed in that direction, there are some circumstances where the price of acting prematurely may be too high.

Genetic Foreknowledge

The struggle to react *a priori* has by no means been limited to human behavior. In modern times it has also intruded into human physiology—a twist Darwin himself could never have imagined.

In 2003, thirteen years after the US Office of Biological and Environmental Research launched the Human Genome Project (HGP), the first human genome was published: none other than the DNA sequence of geneticist Craig Venter. Among the tendencies Venter inherited: he is prone to wet earwax, Alzheimer's, cardiovascular disease, and antisocial behavior.

If there was ever any doubt as to whether humankind has entered a time when future outcomes can be manipulated, the fact that we can now identify our proclivity for specific afflictions should put that question to rest. We stand on the verge of being able to mitigate, delay, and in some cases neutralize the undesirable genetic programming we were born with—in much the same way Venter can use foresight to take steps to guard against any eventuality of cardiovascular disease. The fact is, genomic science has given rise to a plethora of prophylactic measures that can be taken long before there are signs of trouble—a priceless advantage over curing an affliction after the fact. From sickle cell anemia to cystic fibrosis to muscular dystrophy to Huntington's disease and many cancers, scientists are rapidly unlocking the genetic basis for disorders so they can act preventatively.

Yet even as recently as a decade ago, the idea of sequencing every person's DNA and using gene therapy to avert illnesses was unimaginable. Our foresight was limited to what we could stitch together about our family history. If one of our parents died from heart disease, the statistics of experiencing similar problems sometime during our lifetime were greater than people who had no family history of heart problems. And if both our parents suffered the same disease, the odds shot up again. Then, later, we began adding behaviors and other external factors to the mix: whether we smoked, were overweight, lived a stressful lifestyle, resided in an unhealthy environment. These things increased or decreased our odds, though no one knew exactly by how much, or why.

Author of the bestselling book *The Zone*, Barry Sears never had the opportunity to have his genes sequenced, but based on what he knew about his family, Sears knew to worry:

> A sword of Damocles hangs over my head, something I've known since my early twenties. You see, I'm a walking genetic time bomb. I'm genetically programmed by nature to die of heart disease within the next ten years. My early death seems all but inevitable: my grandfather, father and every one of my three uncles were killed by heart attacks before they reached the age of fifty-four. As I write this I am forty-seven.

Though knowing the likelihood of certain diseases is helpful, until the arrival of DNA sequencing there was no way to determine whether we actually had genes which predisposed us to any particular condition. For folks like Barry Sears, there was only enough information to assume the worst and err on the safe side—which meant regular checkups; staying away from sugars, fats, and carbs; exercise; keeping his cholesterol down; and so on. So he did what he could.

Fortunately for Sears, it worked. This year he celebrates his seventieth birthday.

I mention Sears because I suffer a similar dilemma—both my mother and father battled and died from colon cancer. My mother's cancer was discovered late. Before she manifested symptoms, the cancer had metastasized in her liver and other organs. My father escaped my mother's fate by regularly having polyps (the precursors to colon cancer) harvested beginning at age fifty. And though I have yet to have my genes sequenced, the likelihood that Mother Nature has designed me to pass from colon cancer is more than twice the general public.

I live with this fact every day.

While it is true that we will soon have the ability to edit the gene responsible for colon cancer, as of today, regular colonoscopies, a daily dose of aspirin, and a high-fiber diet are the only measures I can take. So I do. Like Sears, I am acting to avert cancer without knowing whether I am in any real danger, or whether what I am doing will have any impact.

But let's say the day comes when I can buy an over-the-counter genetic test and discover my fears are justified. Having the genetic predisposition is not a *sentence for cancer*. In Venter's case, a predisposition for antisocial behavior did not mean he would become a criminal or loner. In fact, the opposite occurred. He went on to become one of the greatest geneticists the world has known, and get married, have a family, develop meaningful friendships, and so on. Nor has he had any signs of heart failure, Alzheimer's, or other predispositions he inherited. Simply *having* a genetic predisposition to a disease doesn't mean we will contract it. It's more complicated.

One way to understand our individual genome is as a panel of "switches" we inherit. Some of these switches are in the "on" position when we are born. For example, our hair and eye color and our facial features are already in the "on" position. Bodily functions such as temperature regulation and heartbeat, as well as behaviors such as the instinct to suckle our mother's breast, cry out when we are hungry, and stand upright are also activated at birth.

Likewise, as we mature, switches that dictate behaviors such as speech, walking, and reproduction all automatically turn on in healthy humans And while humans share the instructions common to our species, we also inherit programming that is unique to each of us. Some of these unique genetic instructions appear to be *time-activated*, while others are triggered by environment and behavior. Take a person's height for example. We might possess instructions to grow tall, but without proper nutrition we may never realize that potential. Similarly, an individual may smoke two or three packs of cigarettes a day and never contract lung cancer—whereas a person who never smoked might die of the disease. Twins may each have the genes associated with breast cancer (*BRCA1* and *BRCA2*), yet only one may actually contract the disease. More and more we are learning the risk of breast cancer may have more to do with how a person's genetic material interacts with the environment. And though we do not understand why two people with the same genetic profile respond differently, today's advanced analytic systems are unraveling the complex interrelationship between the instructions we are born with and our fate. Within the last eleven months we have identified 133 genes responsible for retinal disease alone. We also isolated genetic mutations that make us susceptible to tuberculosis and stuttering. Stuttering!

A couple of months ago I learned that scientists have discovered people who consume sugary beverages such as sodas are twice as likely to possess genetic predispositions for obesity. The following day an article in the *Los Angeles Times* suggested that those who aren't successful may be genetically predisposed to failure:

> Lower socioeconomic status is associated with a range of self-defeating behaviors, including more risk-taking (not using seat belts, for example), worse adherence to protocols (such as failing to complete a full course of a medicine) and poorer financial management (impulse buying, for example, or buying on credit, which adds considerably to an item's cost).
>
> Why is this? One obvious explanation might be that those cognitive traits are what gave rise to poverty in the first place.

Cognitive traits? You mean our genes maybe the impetus behind a life of poverty? And if this is true, are there also genes that predispose us to wealth? Achievement? Fame? Leadership?

As scientists close in on gene expression, the day is fast approaching when every newborn will leave the hospital with more than a knitted cap and blanket. They will come equipped with their own sequenced DNA and instruction manual. And with that, the power to alter their destiny. They can *predapt*.

"The reasonable man adapts himself
to the world; the unreasonable one persists
in trying to adapt the world to himself.
Therefore all progress depends on
the unreasonable man."

- George Bernard Shaw

CHAPTER FIVE

Predaptation

My dog is testing me. As soon as the suitcase comes down from the hall closet, he knows I'm leaving—and not just for an hour or two...

His first test was to lie next to the suitcase while I packed—pleading through the tops of his eyes.

Having failed to alter my plans (the dog-sitter arrived, I loaded the suitcase into the car and drove away) he moved on to Plan B: placing a toy next to the suitcase... just in case I might like to take it, and him, with me.

Again, he achieved no result.

Over time his behavior graduated to going to the farthest room in the house and shunning me when the suitcase came out. This was followed by lying across the threshold of the door leading to the garage so I had to step over him. He would not move. Not for treats. Not for the tennis ball. Not for anything.

So I—still gifted with a slightly larger brain than my dog—adapted instead. I began using the front door. I would wheel my bag all the way around the house to the garage. Then walk back around again, pet him, say goodbye, and leave.

Things could have continued this way, except that I have the benefit of foresight and knew winter was coming. Rolling the suitcase around the house wasn't going to be as easy. I needed another plan.

One afternoon, while I was watching the dog attack the sprinklers in the backyard, I devised a practical solution: First, I would hide the fact I was leaving for as long as possible—no more leisurely packing. Just before I left, I would put the dog outside, turn on the sprinklers to distract him, run in, get the suitcase down, pack, and put it in the car. Then I would let the dog back in.

It was deceitful, but it worked.

My other idea was to "recondition" my dog. But this required time. I began taking him with me when we were not headed to the beach, park, or river. After all, whose fault was it he thought I was headed to one of those places every time I drove away? If those are the only places I ever take him in the car—welllllll?

So now we go to the dry cleaners, grocery store, and post office together, and we wait while I'm having the oil changed in the car. I often glance over at him sitting next to me and he looks as bored as I am.

By supplementing his brain with alternative scenarios—scenarios he has no ability to construct on his own—my dog can now imagine me reading an old magazine at Jiffy Lube rather than having fun at the beach without him.

And one more thing.

While I was coming up with these workarounds, I had another thought: I can head off this problem with any future dog. The next pup only rides when we go to the vet!

While Recorded Future was predicting trouble in Yemen, CAD systems were busy making products safer, genomics was prolonging human life, and Fuzzy Logix was building predictive models to avert diabetes and opioid addiction—I was busy using foresight to preempt my dog.

Okay, so my revelations weren't quite as profound...

On the other hand, this is the nature of evolution. It happens unevenly. Everyone didn't stand upright or discover fire and the wheel at the exact same moment. Nor was everyone equally proficient. Whether it's physical or social evolution, *adaptation is not uniform*. A little here, a little there, then, over time, a group, a society, a species, and so on.

I don't have a Big Data system. And I don't have access to the kind of sophisticated algorithms used by Wall Street or the CIA. One day I will. But not today. So, I use the foresight I was born with to figure out how much wine I need for my dinner party, do some year-end tax planning, and schedule a colonoscopy. In this way, I try to get out ahead of life—even if it's just a little.

By doing these simple things, I am trying to make my future a little easier, a little safer, a little friendlier. I'm preparing for what could happen—or trying to make what I want to have happen a reality. In other words, I'm *predapting*: preparing for or manipulating the future to increase my opportunities for success.

Preparation or Predaptation?

So what's the big deal? What's the difference between preparing for the future and predaptation? Haven't we always looked ahead? Early civilizations were known to make preparations for famine, drought, war, and other threats. Why is predaptation any different?

Good question.

For thousands of years, humans have readied themselves for emergencies and threats—from fashioning spears to hoarding food, gold, medicines. Presently, organizations like the American Red Cross, UNICEF, FEMA, and the World Health Organization prepare us for what may lie ahead. But make no mistake, these organizations do not attempt to *prevent* adverse events from occurring. They merely get us ready for disasters, and step in afterward to deal with the impact. Predaptation reaches past preparedness by eradicating a threat beforehand. There is a vast difference between building more reservoirs to store water and acting to stop the climate change responsible for drought. A vast difference between abortion and practicing birth control. A vast difference between fixing my credit score today and trying to deal with my predicament in the middle of escrow. One is after the fact. The other is before. One is reactive. The other drives an arrow straight through the heart of causation. One bends to change. The other is the change.

Five Tons of Gold

One of my favorite stories of adapting and predapting occurred in 1869. The financial institution we know today as Union Bank began 153 years ago as the Bank of California. The Bank of California has the distinction of being the first "commercial" bank established in the United States.

But within five years of opening its doors, the institution found itself caught between depositors and the federal government in a dangerous quandary from which there appeared no escape. The nation's first foray into commercial banking was destined go down in history as an epic misadventure...

Here's what happened.

In the 1800s, it was common for banks to store large reserves of gold bullion in their vaults. Banks would transfer the raw bullion to a nearby government mint when they needed coins to satisfy customer withdrawals. Then the US Treasury would mint the coins and return the same weight in gold back to the banks in the form of official government-issued currency the bank was permitted to give its customers.

So the Bank of California established its headquarters as close to the San Francisco Mint as possible—a minting facility the US government opened in 1854 to better serve Gold Rush prospectors. But when the Treasury announced they were closing the San Francisco Mint, banks in the West faced a serious coin shortage. They had plenty of bullion, but no fast way to convert it to coin. In those days it would have taken months to transport gold and coins back and forth across the country to other minting facilities.

To head off the disaster, several banks joined together and made a special appeal to the Treasury to allow their bullion to be exchanged for a portion of the $14 million in coins which were already minted, sitting in the nation's capital. But under the order of President Grant, the Treasury turned down the emergency request—thereby setting the commercial banking industry up for an unrecoverable disaster.

Late one evening, Bank of California founder William Ralston confided in financier and friend Asbury Harpending that if they did not have $1 million in coins the next morning, there would be trouble. Then Ralston asked Harpending to do him a favor: meet him at the bank at one o'clock in the morning and tell no one.

What occurred next is largely known through Harpending's journals:

> We walked noiselessly to the United States Sub-Treasury, then located on Montgomery between California and Sacramento streets... A dim light was burning within. Mr. Ralston asked us to halt a few paces from the entrance; then to our great surprise he opened the door to the Sub-Treasury, without challenge of any kind, and closed it after him as he stepped inside. Presently he emerged with several sacks of coin. "Take that to the bank," he said. "The gentleman there will give you something to bring back."
>
> The party at the bank received the cash, tallied it and handed us gold bars for the same value. These we took to the Sub-Treasury, where we found Mr. Ralston smilingly awaiting us with a new cargo of sacks on the sidewalk. We turned over the bars [of gold] and made another journey to the bank.
>
> Thus, at dead of night, passing to and fro, we transferred in actual weight, between the Sub-Treasury and the bank nearly five tons of gold.

Five tons of gold, in sacks, hand carried through the streets of San Francisco at 1:00 a.m. back and forth by a few civilians? Hard to imagine.

Though Ralston found a way to mitigate the coming calamity, it's what he did next that saved the fate of commercial banking.

He heard worrisome rumors were spreading that the banks did not have sufficient coins. It wasn't going to take much to trigger a stampede of withdrawals. So Ralston made a plan. First, he put extra tellers at the windows so there would be no lines. No customer would have to wait to get their money. Then he ordered bank employees to bring tray after tray of gold coins out of the vault and stack them in plain sight behind the tellers.

Seeing there was an endless supply of coins, no lines, no emergency, no cause to worry, customers left without making a withdrawal.

What did Ralston know that others didn't?

He knew that if the environment in the bank was normal, bank customers would behave normally. He might not have been able to stop the federal government from manufacturing a coin shortage. He could do nothing about the closing of the mint or President Grant's refusal to help banks in the West. All he could do was adapt creatively and swiftly to these new conditions. But by putting five tons of gold coins in plain sight, and making sure dozens of cheerful tellers had no problem filling customer requests, Ralston also predapted: he *stopped a crisis from ever coming to fruition.*

And that is how foresight and predaptation saved banking in 1869.

Today, this powerful duo can be found everywhere—affecting how businesses innovate and invest, institutions regulate, and you and I make decisions. But this is different from how humans have adapted in the past. And how other life-forms respond to change.

In nature, adaptation is random. So by definition, evolution is *reactive*—never *proactive*. When a change in the environment occurs, we possess traits that favor that change, or our physiology and behaviors must adjust—within a prescribed period of time—to accommodate the new requirements.

But in modern times, it is not sufficient for an organism, business, government, or individual to react to change. It is not enough to find out we have a genetically transmitted disease after symptoms appear. Not enough to punish a nation with sanctions after a cyber attack. Not enough to adopt a new technology after we learn our competitors have. A vast new universe of foreknowledge has put a twist on adaptation.

The Principles of Adaptation

So how *do* we become better, faster adapters? More effective predapters? How can we avoid the pitfalls that plagued Blitz USA, the subprime mortgage industry, the NFL?

For many centuries, scholars have been studying adaptive strategies in the natural world—those that work as well as those that produce extinction. By observing how organisms thrive in the face of adversity, they've extracted time-tested, universal principles that are essential to success.

Alfred Marshall, Joseph Schumpeter, Karl Marx, and Thomas Jefferson were among the first to apply these principles to modern economics and governance. In the 1800's, economist Alfred Marshall shocked the world by observing parallel forces in evolution and economics. In his book *Principles of Economics* (1890)—a foundational economics text for over a century—Marshall went so far as to borrow from Darwin's use of the expression, "*Natura non*

facit saltum" (nature rarely makes a leap) to explain gradualism in financial markets.

Soon other social scientists followed Marshall's lead—
including Joseph Schumpeter, who took great pains to illus-
trate how evolution operates in capitalism:

> The opening up of new markets, foreign or
> domestic, and the organizational develop-
> ment from the craft shop and factory to
> such concerns as US Steel illustrate the same
> process of industrial mutation—if I may
> use the biological term—that incessantly
> revolutionizes the economic structure *from
> within*, incessantly destroying the old one,
> incessantly creating a new one.

Schumpeter is credited with the idea of "creative destruction"—
a process whereby entrepreneurs disrupt convention, forcing
previous methods, products, and institutions to adapt, mutate,
or go the way of the dinosaur.

And when it comes to governance, Thomas Jefferson
expressed strong views on how governments must evolve—an
idea that was considered provocative and dangerous during
his time. The suggestion that leaders—many of whom considered themselves God-anointed—must adapt to changing
circumstances bordered on heretical. But Jefferson, an early
proponent of democracy, argued that even a foundational doctrine such as the US Constitution must accommodate change:

> Whatever be the Constitution, great care must
> be taken to provide a mode of amendment

> when experience or change of circumstances
> shall have manifested that any part of it is
> *unadapted* to the good of the nation.

Jefferson surmised that, in its original state, the US Constitution had a shelf life of around thirty years. Accordingly, the founding fathers laid out specific procedures for how the document could be adapted. And since 1791, the Constitution has been amended twenty-seven times.

Though we don't often stop to acknowledge it, modern economics, political science, and business *all* take their lead from the same principles that have safeguarded life on Earth for nearly four billion years. With this in mind, it behooves us to understand how the twelve foundational principles of adaptation work in the natural world and modern life.

Principle One
Failure is more common than success.

Scientists estimate that the overwhelming majority of living organisms which once inhabited the Earth are now extinct. Likewise, the number of businesses that shut their doors every year far outnumber those that succeed. And new pharmaceuticals, rock bands, and patents? Many more expire than survive.

Natural selection has no compassion, no affection for justice. It makes no allowance for favoritism, forgiveness, or future regret. Its faithful bedfellow is, and will forever remain, *change*.

The first principle of adaptation requires us to acknowledge that we are operating in a "high failure-rate environment"— one where the number of wrong options exceed the number of right ones. This is the reason picking the right stocks to invest in feels similar to picking winning lottery numbers. The same goes for choosing the right insurance plan, mate, college major, and medical treatment—never mind the correct light bulb at the hardware store.

The odds are against us.

One day Big Data, predictive analytics, and artificial intelligence systems will help us identify the optimal solution out of a sea of inferior alternatives. And digital assistants like Apple's Siri, Amazon's Alexa, and Microsoft's Cortana will vet our choices in less time than it takes to formulate a question. But until they can move past simple search functions and manage complex decision-making, our only option is to embrace strategies which allow us to succeed in spite of poor odds.

A good place to start to turn those odds is to make a distinction between good and bad failure. It turns out, failure is a lot like cholesterol. Too much bad cholesterol will kill us. But we need a certain amount of good cholesterol to keep our bodies functioning properly. The trick is to keep the good cholesterol while lowering the bad. Similarly, not all failure is undesirable. Some varieties of failure are vital to progress.

Vital to progress? Yes, *vital.*

Consider the achievements of Einstein, da Vinci, the Wright brothers—and more recently Hawking, Jobs, Venter, and Disney. They all came at the expense of failure. Henry Ford went broke six times, and Ray Kroc, the founder of McDonald's, more times than Ford. Akio Morita's first product—a simple rice cooker—burned so much rice he sold fewer than one hundred units. Who knew that would be the start of electronics and entertainment giant Sony? Bill Gates' first venture, Traf-O-Data, went down in flames. Dan Brown's first two books received little attention until he penned *The Da Vinci Code*. And many people are shocked to learn that van Gogh sold only one painting while he was alive.

By all accounts, these endeavors were riddled with defeat. Which begs the question: in a "high failure-rate environment" such as science, business, publishing, art, or governance—is there any way to separate good failure from wasted effort? Any way to know which failures to inspire and which to quash?

As a matter of fact, there is.

Dr. Jamer Hunt teaches social and cultural anthropology at The New School in New York City and is an expert on failure. According to Hunt, failure isn't black and white—it's a spectrum comprised of six categories that bleed into one another. In brief, here's Hunt's spectrum of failure from "most devastating to most productive:"

- **Abject Failure:** This is really the dark one. It marks you and you may not ever fully recover from it. People lose their lives, jobs, respect, or livelihoods. Examples: British Petroleum's Gulf spill; mortgage-backed securities.

- **Structural Failure:** It cuts—deeply—but doesn't permanently cripple your identity or enterprise. Examples: Apple iPhone 4's antenna; Windows Vista.

- **Glorious Failure:** Going out in a botched but beautiful blaze of glory—catastrophic but exhilarating. Example: Jamaican bobsled team.

- **Common Failure:** Everyday instances of screwing up that are not too difficult to recover from. The apology was invented for this category. Examples: oversleeping and missing a meeting at work; forgetting to pick up your kids from school; overcooking the tuna.

- **Version Failure:** Small failures that lead to incremental but meaningful improvements over time. Examples: Linux operating system; evolution.

- **Predicted Failure:** Failure as an essential part of a process that allows you to see what it is you really need to do more clearly because of the shortcomings. Example: the prototype—only by creating imperfect early versions of it can you learn what's necessary to refine it.

Hunt makes the point that *version* and *predicted* failure are necessary for progress because these forms of failure facilitate testing and learning. Without them, innovation would be impossible. But other types of failure? Hunt observes, "There's no question that out of failure—even 'abject failure'—we emerge transformed in ways that may ultimately be beneficial. But that does not mean all failures deserve a trophy."

There is failure that warrants advocacy, and failure that must be expelled, and it takes a wise leader to know the difference.

Fail Fast!

Over the past thirty years, I have worked directly for and consulted with many of the biggest conglomerates in the world—organizations that say they want to be more innovative, move faster, and embrace disruption, risk, and failure. They say they want to be "first movers"—leap ahead of competitors and grab market share. They say they want to act more like a start-up, while continuing to service the brands responsible for their past success. They all say it. And pay big money to have experts tell and show them how. Then a majority turn right around and get busy reducing everything into the most efficient process possible—quashing and punishing failure anywhere they find it. Which, in a high failure-rate environment, is the kiss of death. Because *failure cannot be avoided in a high failure-rate environment.*

No company taught me more about embracing failure than Dole Fresh Vegetables. And no executive ever had a better handle on how to build a forward-facing culture than President Ray di Riggi.

By the time I was introduced to di Riggi, agriculture was already undergoing a revolution not seen since man's first domestication of plants. A business that once consisted of putting a seed in the ground, watering, and harvesting, had morphed into something that had more in common with a pharma laboratory. From genetically modified foods, nutrient-fortified fruits and vegetables, microencapsulation, and the hunt for good bacteria, to inventing synthetic fertilizers and hybrids, the job of feeding the world was growing more complex by the minute.

But di Riggi was unfazed. He pushed Dole to the brink—forcing the largest grower in the world to reach new efficiencies. Right down to reducing the number of color copies printed, di Riggi cut waste to the core, streamlined processes, and automated wherever he could.

But there were signs the gains were beginning to slow. Vice President of Human Resources Kent Hansen observed:

> During the first three years, the Operational Excellence (OE) program lowered the absolute *cost per case* by roughly 10 percent. As I recall, about 10 percent of that savings came in year one. Almost two-thirds in year two. And the remainder in year three. But year three was significantly more difficult. And looking ahead, we could see it was going to get harder."

The big cost-cutting measures had been identified and acted on, so what remained were smaller, incremental improvements. And they would not be enough to sustain Dole's momentum.

What was next? What could Dole do to get out ahead of the market? How could the company increase their appetite for risk innovation? For real innovation? These were the questions di Riggi put before his team.

Di Riggi challenged Dole executives to take an outward-facing view of their industry. He encouraged them to invest in projects aimed at disrupting traditional agriculture methods and thinking. And he made sure they knew that failure was not only necessary, it was expected.

Di Riggi laid out three simple guidelines for new ventures: 1) clearly define—in as much detail as possible—the metrics to track progress before starting, 2) meter out the financial investment based on achieving specific milestones (Round 1, Round 2, and so on) and, 3) failure should be contained so successful areas of the business would not be affected—until, that is, a concept had been proven.

Then di Riggi instructed his reports to "fail fast!" By failing quickly, resources could be reapplied to other projects that showed promise.

And fail they did.

As the months passed, some ventures proved auspicious. Others proved success was unlikely and were killed. Still others required more time and capital than were worthwhile. And true to form, many failures begat new ideas no one had previously considered: "From the ashes a fire shall be woken, a light from the shadows shall spring."

Eleven months after di Riggi said "fail fast," Dole's ship came in. One of the highlights of their annual company meeting was a crude, field-shot video of a lettuce harvester—a machine experts said *could never be built.*

Lettuce is not only a delicate, time-sensitive product, the margins are miniscule. It doesn't take much to damage the leaves. And a case of brown lettuce quickly ends up in the trash bin behind the grocery store. It doesn't take much to wipe out the profits of an entire field if harvesting and transport isn't treated with kid gloves—which means picking and handling by human hands.

Making automation more challenging is the fact that no two heads of lettuce come out of the ground the same way, or are the same size. Cut the stem too high and you wind up with a handful of leaves. Cut too low and you're stuck with a long, unsightly stalk. An automated harvester would have to know exactly where to cut each head, then gently pick it up, place it on a conveyor belt, carefully wash it, and gently lay it in a bin. And do it cheaper than manual methods could.

That said, labor costs were on a steep incline. New immigration policies had created a dangerous shortage of pickers in the United States—a shortage which threatened how fast crops could be pulled from the field. At the exact same moment Dole's lettuce crops were ready for harvest, every other grower's crops were also ready. So, every agriculture company found itself competing for the same dwindling labor pool—a pool that had become so small entire fields were going to waste.

An automated lettuce harvester would solve these and other problems.

So Dennis Castillo, Serafin Albarran, and Steve Jens set their sights on the impossible: a mechanized harvester that could be run by two or three operators.

Their first attempts were disheartening. The early prototypes would have made better rototillers. Photographs of mangled lettuce and torn-up fields where machines had pillaged crops were difficult to look at. But Carlos Meza, manager of Lettuce Services, was just getting warmed up. With each failure, the team was learning. And now that di Riggi had made "fail fast" the order of the day, Meza, Castillo, Albarran, and Jens faithfully reported on every failed attempt at the executive staff meetings—often breaking into laughter as the team showed video footage of iceberg lettuce flying like basketballs through the air. But the most impressive thing that occurred during these meetings was that di Riggi and other executives smiled right along with them. Di Riggi missed no opportunity to praise the team for their persistence and remind them that they were getting closer.

Bear in mind, no one knew how the venture would turn out. There was no way to know if they would *ever* develop an operational unit. And if they did get one to work, would it be practical to operate? Would it be durable and portable and meet food and safety regulations? And what would it cost? Would it be cheaper than hand picking? There were many unanswered questions. And no way to answer them in advance. All the team could do was inch forward, one error, one small triumph, at a time.

Today, Dole's patented lettuce harvester has eliminated over 60 percent of the labor costs associated with bringing a lettuce crop in—and it does it faster and more consistently than manual harvesting can. As a result, Dole is able to offer a better product at a lower price than their competitors. What's more, the company can plant larger fields because lettuce can be brought to market before the picking period has lapsed. As if those benefits aren't enough, as soon as the harvester proved viable for lettuce, the proud parents couldn't wait to repurpose their breakthrough for other crops.

So, the next time you dig into a salad, remember the years and months of failure that led to that delicious meal. Then fail often! Fail fast!

Principle Two
The greater the magnitude, speed, and complexity of change, the higher the rate of failure.

The second principle of adaptation is related to the first: expect greater rates of failure when change is severe, rapid, or multifaceted.

The Ediacaran Period—a time when single-celled organisms were just becoming more sophisticated life-forms—is an example of the second principle at work. During this period, water-dwelling organisms depended on oxygen in the same way you and I do. But as oxygen levels began quickly dropping, more than half of these organisms were unable to adapt. And those that failed became the fuel that powers our cars and heats our homes today.

Many people are also familiar with the mass extinction brought on by the Ice Age. Fossils and other evidence show that cold-adapted creatures moved *en masse* to lower latitudes, whereas animals that required warmer conditions failed to survive. Scientists speculate that 60 to 80 percent of life succumbed to this sudden, pervasive shift in temperature.

More recently we observed this principle at work during the subprime mortgage collapse, when the real estate prices began plummeting faster than financial institutions could adapt. As the contagion picked up velocity, the US government had no choice but to intervene with the emergency Troubled Asset Relief Program (TARP)—a program aimed at buying time for lenders to make necessary adjustments.

But it wasn't only the speed at which the mortgage market collapsed. The collapse was also magnified by complexity. As much as folks prefer to blame legislators, greedy mortgage brokers, and unqualified homeowners—the root of the problem was convoluted derivatives and credit default swaps which even the most capable financial experts did not fully understand. They may have been buying, trading, and valuing these instruments, but testimony later revealed the industry had no idea what the basis for the assets really was.

And then there was the extent of the contagion. Within two days, every country's economy was affected—many of which had never heard of a subprime mortgage and had no connection to the US real estate market.

It doesn't matter whether an abrupt change is physical, social, or otherwise—time, severity, and complexity conspire to make successful adaptation difficult.

Principle Three
Any drive toward singularity is a drive toward extinction.

As of today, scientists have classified more than 12,500 species of ants. And there are twice as many species of fish as there are ants. Never mind the numbers of birds, beetles, and butterflies. You might ask yourself, why so many? Why not just one variety of ant, bird, or fish?

Why are some humans tall and others short? Some inclined to be fat, while others eat everything in sight and never gain a pound? Why are some hairy, big-breasted, flat-footed, long-toothed (hopefully not at the same time)?

The answer is, for the same reason we don't invest all our money in a single stock. The natural world works the way investments do. Nature *spreads the risk around by relying on randomness.* Which means when conditions change, some investments pay off, while others don't. Diversification acts as an insurance against unilateral failure. In any high failure-rate environment, singularity is the enemy of success.

Take the euro for instance.

No matter how many times economists explained why the euro was a good idea, I objected to it from the start. Moving to a single currency flew in the face of the third principle of adaptation. With this in mind, I forecasted the euro—along with the move to relax borders between nations which adopted the euro—would cause participants to become unstable and vulnerable.

Unfortunately, no one was looking to a sociobiologist to gain a better understanding. No one was concerned about the relationship between diversification and sustainability. Or the principles from which the very fabric of modern economics were derived. So in 1995, seventeen out of twenty-eight countries comprising the eurozone adopted the euro as their official currency. Shortly afterward, more countries followed suit, and many African nations also tied their fate to the new currency.

As of 2017, there were an estimated one trillion euros in circulation, propelling it to become the second-most traded currency in the world—and, from my vantage point, bringing the world ever closer to financial collapse than any other economic initiative in recent years.

But the danger did not start with the euro. As far back as the 1992 Maastricht Treaty—the impetus for the euro—the drive toward singularity was rearing its head. The treaty spelled out the requirements each nation must meet to adopt a single currency. One of those requirements was that each country must maintain a deficit of less than 3 percent, and a debt ratio of less than 60 percent of the country's gross domestic product (GDP). Euro participants were also obligated to sustain low levels of inflation and maintain interest rates that were "equivalent to other eurozone countries."

After spending time examining the Maastricht Treaty, my first question to government leaders was *What happens if they don't?* What happens if one country mismanages their economy? Their debt swings wildly out of control? They fall prey to hyperinflation or worse? Given volatile global markets, it wasn't difficult to imagine any of these real possibilities.

By 2009, seven years after the euro was adopted, a few participants began to make noises about treaty requirements. Unbeknownst to other eurozone nations, countries like Greece embarked on a convoluted plan to mask their financial problems. In addition to adopting unconventional accounting practices and allowing off-balance sheet transactions, they invented a new class of derivatives based on the future income of Greece. Rather than issue more bonds—which the government would be obligated to pay back and would have increased their indebtedness—this new class of derivatives was an *investment based* on "an opportunity to share in the future income" of the Greek government. The complicated scheme gave the appearance that Greece remained in compliance with the treaty—at least for a short time.

By 2010, other eurozone nations demanded that Greece's finances be made transparent so the extent of the deception could be determined. Were it not for the fact Ireland, Portugal, Spain, and Italy were also struggling and engaging in creative accounting schemes, the Greek economy may well have been left by the side of the road to fend for itself.

As the insolvency spread, the euro, and every nation that put their faith in it, found themselves in trouble. The train had reached the end of the tracks.

This is what happens when diversification is abandoned in favor of singularity.

Having no other choice, the International Monetary Fund, Eurogroup, and European Central Bank jumped in with a €110 billion loan to Greece, followed by another €130 billion to keep Greece's mismanagement from taking every other country down with them. A rescue committee called "the troika" was formed, and soon new regulations and oversight organizations were put into place (such as the European Financial Stability Facility and the European Stability Mechanism). The European Central Bank pitched in by dropping interest rates and opening up more credit. Other eurozone countries were also offered bailout programs and put on an austerity diet. As of 2017, the situation has stabilized. And thanks to advances in Big Data and predictive analytics, compliance issues in the future can be dealt with long before they become problematic.

But here's the million-dollar question: how would each eurozone nation have fared had their currencies remained independent? Had they remained *diversified?* Based on what we know about how diversity works in nature and high failure-rate environments, we can now answer that question: some countries would have done better, some would have not been affected at all, and economies, which failed to adapt successfully, like Greece, would have imploded.

I was not in favor of the move to the euro in 1995, and I remain opposed to it for the same reason today. For the same reason the world's financial markets were better off prior to World War II, when economies were less intertwined, less enmeshed, less homogeneous. At that time, the shenanigans of one country did not have the dangerous domino affect it has today.

Want proof?

Most of us have forgotten that in 1944, as World War II was coming to an end, Greece found itself in a much bigger, deeper hole than it's digging out of today. Greece fell victim to unstoppable hyperinflation. Every few days, monetary values began dropping by half. The country's national income fell 70 percent and foreign trade came to a virtual standstill. Yet, one of the worst cases of hyperinflation in European history, had a negligible impact on nearby nations. Not even close to the threat posed to euro participants today. Which serves to make the point that any effort to decrease diversification—to diminish variety—is a formula for trouble. Once we get down to one type of ant, one type of elephant, tiger, or bear, the writing is on the wall. It's all a matter of time.

Principle Four
Success warrants imitation.

There's a reason most of us can't tell the difference between a viceroy butterfly and a monarch. Monarchs have an unpleasant taste that birds avoid. The viceroy doesn't come equipped with this defense, so they do the next best thing. They mimic the appearance of the monarch—to the point of being virtually indistinguishable by birds or humans.

The syrphid fly adopted a similar strategy. It disguises itself to look like a yellow jacket despite having no stinger.

Similarly, Puff Daddy and Eminem hit the stage wearing jeans falling off their torsos and baggy T-shirts, and the next day every wannabe rapper was rushing to the mall. Once Apple made icon-based computing popular, everyone jumped on board. And there's a reason every television looks exactly the same.

Imitating success works. It is more efficient, less risky, and easier. Which is why there are more copycats than innovators in every field of endeavor.

Similarly, we already have successful models for most of the challenges humankind faces. For decades, the Japanese public education system has outperformed other countries. *Bloomberg News* ranks the healthcare systems of Hong Kong and Singapore among the most cost-effective and advanced in the world. Kyocera, ABB, and Pfizer are among the best-managed companies. Switzerland and Singapore continue to have the lowest crime rates in the world. And according to the Worldwide Governance Indicators project—which tracks the effectiveness of governments—Finland, Denmark, and Sweden administer the same services other governments do, but do it better. Their citizens consistently garner the highest scores for happiness.

Yet how often do we look for models to emulate outside our country? Our industry? Our discipline? What can an economist learn from a biologist? A politician from a physicist? A produce grower from a hospital?

It turns out, plenty.

Parallel Paradigms

After being invited to observe Dole's remarkable transformation, I was eager to see if I could find successful models of adaptation in other industries which Dole could adopt to move even faster. But where to look, where to look...

One afternoon, while listening to the executive team explore strategies for reducing the time from "farm to fork," I said something that startled even me: I suggested that *Dole was no longer in the food business.* The industry was experiencing such radical change, Dole had more in common with a hospital emergency room than their farming counterparts.

An ER?

For a second, even di Riggi looked puzzled. But consider the similarities. The moment a head of lettuce is pulled from the ground, a berry plucked from a bush, a banana cut from a tree, *it is dying.* We are thrown into a mad rush to get the patient from the field to the grocery store while it is still alive. Furthermore, the condition of an ER patient's health *before* they are admitted—as well as how they are handled en route to the ER—largely determines their odds of surviving. This is not largely different from what happens to fresh produce. What takes place in the field *before* the plant is harvested—as well as en route to the grocery store—has everything do with whether produce will arrive healthy or bound for the morgue (dumpster behind the grocery store).

In effect, Dole was an *über* hospital—processing millions of live patients, arriving daily from every part of the world. The company's job was to extend those patients' lives for as long as possible. From this perspective, Dole faced the same mission-critical, time-sensitive challenges today's ERs face. So why not adopt the procedures, technologies, protocols, language, and policies emergency rooms use?

This idea sent Dole Fresh Vegetables in a new direction. They suddenly saw themselves as fast-moving ER workers on a mission to save the life of every strawberry, every head of broccoli, and pineapple in the field—proving that, when it comes to adaptation, often the quickest way to predapt is to emulate.

What Not to Reinvent

I credit most of what I learned about imitation from my first job. When I was fifteen, our high school football coach asked if I would be interested in babysitting his children from time to time. He had practice and games on the weekend and his wife worked as a secretary in a busy real estate office. Though she wanted to spend the weekends with her children and husband, this was the time when realtors wrote most of their offers and needed her most.

But after babysitting for a few weeks, my coach and his wife sat me down and asked if I would be interested in switching jobs. How would I like to learn to type contracts and letters and answer the phone in the real estate office on the weekends instead? The job paid two dollars an hour more. I jumped on it. It was a real job.

The realtors set up a small desk next to the break room for me—complete with an IBM Selectric typewriter, which, in those days, was a giant machine that took up half the desk. Next to the typewriter was a stack of carbon and typing paper, some preprinted contracts, a small bottle of Wite-Out for mistakes, and an in-and-out box. Every Saturday I was the first to arrive at the office. I turned the lights on, made coffee, took down messages left on the answering machine, then began going through my in-box—organizing the typing according to what was needed first. When I finished what was in the in-box, I had nothing to do. So I went office to office asking the agents if they needed help. And they always had something.

One afternoon an agent waved me into his office. He pulled his business card out of his wallet and said "See this? It's a mess. I need new cards. See if you can make something that looks better."

I'd never looked closely at a business card before. So I took his card and went back to my desk.

Remember, this was the '70s. I couldn't go online, print out samples to show him, and order from any one of dozens of cheap card sites. Not only weren't there any computers, there was no Internet. So after thinking for a moment, I decided the logical next step was to collect other agents' cards. Which—it turned out—wasn't helpful. All of the cards were different.

When lunch rolled around, I hopped on my bike and pedaled up the block to the print shop to see what they had. But when the fellow working the machine behind the counter handed me a binder filled with business cards and went back to his work, I knew I was in over my head. There were cards of every type—some with logos, some with drawings of people washing windows or carrying ladders, some with printing on both sides, others with more than one color of ink, raised lettering, odd shapes, cutouts, and on and on.

I closed the binder and motioned for him to come over.

The man must have had a daughter my age. Or perhaps he saw my bike propped against the wall outside—the one with the goofy pink basket my dad tied on the front with wire. Maybe it was the fact that I looked overwhelmed, desperate, serious, and scared. More than likely he saw an opportunity to get business from his real estate neighbors up the block.

Whatever the reason, he turned off his printing press, walked over, and began flipping through the binder. Then he pointed.

"That one. We print that one for Xerox."

"Xerox?

"You have any idea how much Xerox paid for that card?" He smiled, "They pay a company to come up with a lot of different designs. Then they test the designs, they change them, and test them again. And that's what they came up with. That one."

I nodded. I was being schooled and I appreciated it.

He pointed to the card again. "If it's good enough for Xerox, it's probably good enough for your boss."

So I asked for a photocopy to take back to the office, where I dutifully explained that Xerox had hired an expensive company to research business cards (repeating the words of the printer as best as I could), and after considerable research, this was the one they picked. "I thought we should make your card like this since we know it already works," I concluded.

The agent nodded, "Sounds good to me."

Victory!

From that moment on, I vowed never to reinvent unless necessary. I would look for successful models—solutions with an established track record. And where I could find no satisfactory remedy, that is where I chose to innovate.

Much later I went on to mimic how successful people dressed, how they spoke, wrote, and behaved in business. If you had asked me if I liked what I was wearing, or the way my hair was cut, or how my business cards looked, I would not have been able to tell you. Whether I liked these things wasn't the point—I was only interested in Did they work? and Were they best? I knew others had invested far more time than I had interest in solving these challenges. The viceroy butterfly strategy worked just fine for me. Still does.

There's a popular saying: "Mimicry is the sincerest form of flattery." But this doesn't come close to addressing the important role imitation plays in a high failure-rate environment. When we adopt proven solutions—when we go so far as to institute reconnaissance teams to go out and search for solutions, processes, and breaking technologies outside of our area of expertise—we seize an incalculable advantage.

Principle Five
The size of an environment determines growth.

Every territory has a limit to how many creatures it can support, just as every industry has a limit as to how many companies can operate profitably, and every city has a limit as to how many residents its roads, water, utility, and sewage systems will safely accommodate.

When it comes to understanding the balance between territory and success, social insects have the upper hand over humans. They never allow their numbers to exceed what the immediate environment can support. Yet, even with all our data, foresight, and powers of preemption, we often turn a blind eye to the fifth principle of adaptation.

Take the failure of solar panel maker Solyndra, for example—a case which demonstrates what happens when more challengers enter a territory than it can support.

Solyndra was founded by technologists who spotted a problem with the way solar panels were being designed: collection cells that directly faced the sun when it was overhead became less efficient at energy collection when the sun was at an angle—which was most of the time. Solyndra had a solution to this problem: collection tubes which optimized energy collection from every angle, and used less expensive materials than traditional panels. The company's patented design, combined with the use of new high-yield photovoltaic cell technology and impressive robotically controlled factory, quickly positioned Solyndra as the new leader in solar energy. At their height, Solyndra had one thousand systems in the field, producing over 100 megawatts of power. The company's revenues skyrocketed from $100 million to $140 million in a single year and backorders were stacking up faster than they could ramp up.

Given Solyndra's early success—and the solid track record of the founders—the US government stepped in with $535 million in loan guarantees to match capital raised from the private sector. The loan was intended to generate more "green" jobs at a time when unemployment figures were on the rise. And overnight, Solyndra became one of the best-financed companies in technology history.

But then something went terribly wrong.

Solyndra planned to drive the price of their panels down as demand and production ramped up. This is how economies of scale work: as manufacturing volumes increase, the cost of goods and production come down. And this economy is reflected in the price.

When the price of the most expensive material used by traditional solar panel makers began falling, China and Taiwan began slashing prices and flooding global markets with more products than there were consumers. In 2008, the price of solar modules took a nose dive from $4.20 per watt to $1.20 per watt—a sudden 71 percent drop. On a good day, Solyndra panels produced energy at $2 per watt—eighty cents more than their Asian adversaries.

Every day, as more ant colonies competed for the same patch of dirt, the situation grew more desperate. A bloody price war broke out to bring supplies in line with demand, but Solyndra was ill equipped to absorb massive losses. Within a few months, it was triage time. On Sept 1, 2011, Solyndra filed for bankruptcy.

An immediate investigation into what happened to the half a billion in government loan guarantees was launched, and predictably, the press went on a witch hunt for someone to blame. US Secretary of Energy Steven Chu was accused of cronyism, a failure to provide oversight, and lack of business experience. Others claimed President Obama's green jobs campaign was a sham. Some accused the founders of misleading investors and fraud. Solyndra became the poster child for government incompetence and everything that was wrong with business.

But blaming Solyndra's fall from grace on an unproductive government investment is like blaming my son's poor grades on his tuition. Solyndra wasn't a bad investment. It was an investment that went bad. There's a difference.

When too many competitors attempt to live off the same market, the correction is always painful—resulting in cannibalization and carnage. And while experts may prefer to use terms like *market* instead of *environment*, what holds true in nature holds true for all ecosystems. Markets are ecosystems. And there is only one thing to do when the resources of the ecosystem are insufficient. Only one thing to do when our slice of the pie is too small to sustain us. We must grow the pie. We must expand territory.

All the Territory in World

In 1867, Karl Marx published *Das Kapital* (*Capital: A Critique of Political Economy*), a book that suggested the success of capitalism depended on two main factors: finding and controlling raw materials and cheap labor, and conquering new territories (consumers) where finished goods could be sold. In Marx's view, without continuous expansion, capitalism would exhaust resources and saturate markets, and the entire system would collapse.

But in the 1800's, resources and markets meant those that were within the reach of horse-driven wagons, ships, and trains. There was no way for Marx to envision how the Internet would allow every manufacturer in the world to compete for every consumer in the world. Nor could he have imagined that China would corner 95 percent of the total rare earth minerals available on the entire *planet*—minerals other nations need to produce computers, weapon systems, and other electronically controlled devices. Basically everything.

Today, every person is attempting to survive off one single territory: the planet itself. This fact has produced more intense competition than businesses and individuals have experienced in human history, and largely changed how we perceive "territory." Large conglomerates now compete with small operations in remote areas—operations that have the same access to consumers, distributors, etc.

If expansion is the engine that drives capitalism, then once global markets for raw materials, cheap labor, and customers have been conquered, there's nowhere to go. So the same competitive dynamics that occur in nature when territory becomes insufficient (natural selection) are bound to surface. The fight for survival gets rough.

It doesn't matter if it's solar panel makers, insect colonies on a remote South Pacific island, or the success of a stalwart institution like capitalism—the fifth principle of adaptation teaches us that we cannot grow any larger than the resources in a territory will allow.

Principle Six
Nature's defenses: quickness, cunning, and camouflage.

The strategies which protect animals from threats are not only ingenious but cover a broad range of instinctive and learned behaviors. When threatened, the horned lizard shoots blood from its eyes. The pangolin is covered in spiked armor from head to toe, which it turns outward when predators approach. And when it comes to the French Guiana termite, evolution has endowed older termites—which would otherwise be useless to the colony—with a bright blue spot on the top of their head that contains exploding crystals. Nature has turned the elderly into suicide bombers who sacrifice themselves for the greater good of the troop.

Though the defense systems animals rely on are diverse, in general they can be grouped into three categories: 1) quickness, 2) cunning, and 3) camouflage.

Take quickness for instance. Owing to their great speed and dexterity, hummingbirds have escaped becoming a primary food source for any known predator. And cunning? Who can ignore the opossum's talent for playing dead? Or the Venus flytrap's seductive leaves, designed to trap unsuspecting insects? And as far as camouflage goes, it is difficult to turn our gaze away from a chameleon as it slowly disappears into the background, perfectly disguising itself from nose to tail. Or the puffer fish whose slow, awkward method of swimming should make it vulnerable to predators, except for the fact that it has an elastic stomach that, at a moment's notice, fills with water, giving the fish the appearance of being bigger and pushing its poisonous pointy spines outward like a porcupine.

Human beings and organizations rely on similar methods to fend off danger.

Take the advantage of reacting quickly. The moment it became possible to keep up with changes in stock, currency, and commodity prices every picosecond, the financial industry shifted to programmed trading. There was no time for a human being to analyze data, draw a conclusion, get approvals, and authorize trades quickly enough—especially in the middle of a meltdown. So programs authorizing machines to sell and buy when certain parameters were met became common—mitigating large losses during sudden declines. The fact is, shortening reaction time has become such an important defense against loss that financial institutions now argue over whose equipment is located closer to the stock exchange's computing systems. The time it takes for electricity to travel one additional foot might mean millions.

And how about cunning? Do we use that to ward off threats? You bet.

Airline mileage programs are a perfect example of how cunning protects us from would-be predators. The purpose of customer loyalty programs is to reward customers for their patronage in order to prevent them from leaving for other offers. But take a close look at how these programs work. Are we really getting airline tickets, hotel rooms, gift cards, and rental cars for *free*?

I think you know the answer...

In reality, we pay more than a ticket is worth so the airline can hold the excess in an account and return it back to us later in the form of overpriced products and services. And though most of us know the items we redeem with our mileage points aren't really free, we like to pretend as if they are. But let's be clear here: *free is when I don't have to give you anything.* If I have to pay you first, it isn't free. If I have to purchase a six-pack of soda to get another six-pack at no cost—not free.

If you ask me for my credit card information beforehand—also not free.

At present, every airline, hotel, and retail company employs some version of refunding overpayments in the form of allegedly "free" premiums. Can you imagine anything more cunning that giving me back what was mine to begin with? But there's no question whether these deceptions keep competitors at bay. Last month I checked my United Airlines statement and I had 135,000 miles—five thousand more and my son and I can go to Japan. No need to guess who I chose for my next business trip.

And when it comes to camouflage, we would have a difficult time distinguishing the sand grasshopper from the beige, brown grains of sand it calls home. The walking stick cleverly blends in with surrounding branches, and were it not for its black nose, even an animal as large as a polar bear would be difficult to spot against the Arctic landscape. Disguises are valuable in nature.

We borrow this strategy from nature, too.

During the fourteenth century, Japanese ninjas began wearing black to camouflage themselves during night missions. It didn't take long before villagers began ascribing supernatural characteristics to this elite class of assassins who seemingly appeared out of thin air and disappeared back into it.

In the eighteenth century, German Jäger rifle units switched to drab brown and green clothing to blend in with their natural surroundings. And today's khaki military uniforms came about when the British Army occupied India and discovered their red tunics attracted gunfire. They quickly switched over to sand-colored shirts and slacks. Camouflage continues to be used by the military to protect soldiers, weapons, transportation units, and high-value facilities.

Camouflage is also a highly effective business strategy. We often hear of start-ups and small companies wanting to "fly under the radar" for as long as possible, fearing an attack by bigger, more resourceful competitors. And many businesses use branding as a form of camouflage. Most consumers have no idea Hidden Valley salad dressing, Kingsford charcoal briquettes, and KC Masterpiece barbeque sauce are manufactured by the Clorox Company. And for good reason.

But to really understand how much we rely on camouflage, we need look no further than the box of cereal on our kitchen table.

In 2008, cereal makers began feeling the effects of global fuel shortages. The price of transporting and manufacturing was rising by the minute, as was the cost of ingredients. So companies like General Mills adapted by doing the simplest thing possible. They reduced the content of cereal inside the box rather than raise prices or make smaller boxes.

Overnight, an 11-ounce box of cereal became an 8.7-ounce box for the same price—a price we had grown accustomed to paying. Though the contents may have been a third less, it sure looked and felt the same. Soon the makers of chips, crackers, and other packaged food products were all reducing contents instead of package size as an adaptive strategy. These days there's more air than chips in a bag of potato chips, and I have yet to open a can of anything that was filled to the brim.

But why pick on business when every one uses disguises to fend off failure? A bald man wears a toupee to avoid rejection, and a woman a padded bra to avoid the same. Shoe lifts, dentures, flashy jewelry, and dresses with geometric designs—these are also effective camouflage.

It turns out, quick reflexes, cunning tactics, and camouflage are just as effective a defense in nature as they are for every discipline, industry, and human endeavor.

Principle Seven
Nature's offense: strike first, strike hard.
It's a matter of life or death.

In nature, offensive strikes are driven by necessity. They are generally associated with tremendous risk, frequently producing a life-and-death struggle. And in the wild, even a minor injury can result in painful, protracted death—so for this reason first strikes are reserved for a) an unavoidable threat, b) hunting, c) mating, d) establishing troop order, and c) acquiring new territory.

Among humans, first strikes are much more common. And also frequently rewarded. If a company surreptitiously learns a competitor is going to beat them to market with a new product or feature, it's common to slash the price of their current products to saturate the market ahead of time. The company that has fallen behind might also prematurely announce, and take advance orders, for their next product in an attempt to undermine their competitor's advantage.

The Boston Tea Party was an advantageous first strike, as was Japan's attack on Pearl Harbor, and the US march into Iraq. And when it comes to offensive strikes, more than one spouse has been caught off-guard by a partner who made secret plans to leave.

Recently, a friend of mine made the difficult decision to leave her marriage. The relationship had been a tumultuous one so the decision had been a long time coming. While I was sad to hear the news, I was more surprised when my friend said that she had enrolled in "divorce university."

Divorce university? Never heard of that before.

At divorce university, spouses who have decided to end their marriages use foresight to preempt what lies ahead. They learn to make copies of all the family's financial documents, such as bank and investment statements, real estate holdings, etc., and hide the copies in a safe place outside the home. Divorce university teaches a person how to open a separate checking account and stash away money so there is something to live off while the divorce is proceeding. Students learn to transfer the car and other belongings into their name; apply for new credit cards and health and other insurances; make plans to have the locks in the house changed; have the phone service transferred and utility bills mailed to a private mailbox, and ask a nearby friend to stand by in case there is trouble when it comes time to disclose their intentions to their partner.

In my friend's case, once her husband realized she had planned the divorce down to the smallest detail—and neighbors were standing by—he packed his bags and left without a single cross word.

While some might find the whole idea of divorce college maniacal, from my friend's perspective, she had averted a potentially volatile situation the same way Ralston usurped a run at the bank, or I had repaired my credit report in advance of applying for a mortgage. She was using her foresight to manipulate a future outcome to her advantage.

Sometimes we strike with guns, tanks and bombs, sometimes we use the law or a clever strategy. When we cannot avoid danger, then we are designed to attack it with our full might, as if our survival depends on it.

And sometimes it does.

Principle Eight
Incremental adjustments pose less risk.

Here the expression "nature rarely makes a leap" takes center stage. The reason evolution involves minute adjustments over a protracted period of time is because large corrections are more painful and prone to fail. The trick is to successfully manage a little change here and a little there, and wherever possible, get out in front to minimize what we must otherwise adapt to.

This is why I am not a fan of the word *disruptive*.

These days the idea of disruption has become very popular in business and governance. It's often described as some integral part of innovation—something forward-facing companies embrace and nurture. But let's be honest. Disruption is only necessary where evolution has failed. There is no need for government coups, bail-outs, sudden lay-offs, crash diets, or bankruptcy when we practice gradualism. Aside from unavoidable acts of God or game-changing breakthroughs in science and technology (which are very few), there is no reason to view disruption as more favorable than making much safer, incremental adjustments along the way. And there is no better example of the eighth principle in action than the present plight of AM radio.

Rush Radio

Prior to Rush Limbaugh hitting the scene in 1984, there was no such thing as talk radio Talk was limited to charismatic DJs who brought us the news, weather, and traffic during morning and afternoon drive time. The rest of AM radio was devoted to music.

Love him or hate him, Limbaugh is to be credited with transforming AM radio into a powerful new medium. And no one has been able to knock Limbaugh from the top spot in radio for over twenty years. Which is surprising because he is also the reason AM radio stations are going broke.

The change-over to talk radio happened fast—violating the eighth principle of evolution, which advocates making incremental adjustments. The disruption occurred for two reasons. First, because the number of AM listeners were on the decline due to increased competition from cable television, the Internet, satellite programming, and mobile devices that made it possible to access content anywhere, anytime. And second, with fewer listeners, advertising revenues were plummeting. So when Sean Hannity, Michael Savage, Laura

Ingraham, Bill O'Reilly, and dozens of others began attracting big audiences, AM station owners were primed to switch horses.

Though Limbaugh and other popular talk show hosts may have recruited large audiences to radio, station owners later discovered these audiences did not translate to advertisers. Sponsors were reluctant to have their brands associated with controversial, partisan content no matter how many listeners tuned in. Why would a company advertise on a program that offended half their prospective customers? It wasn't good business.

Convinced advertisers could not afford to stay away if his numbers continued to climb, Limbaugh grew more emboldened:

- "Slavery built the South. I'm not saying we should bring it back; I'm just saying it had its merits. For one thing, the streets were safer after dark."

- "Feminism was established to allow unattractive women easier access to the mainstream."

- "Let the unskilled jobs that take absolutely no knowledge whatsoever to do—let stupid, unskilled Mexicans do that work."

And the more outrageous Limbaugh grew, the more his show grew in popularity.

But there was another side to the popularity. It didn't take long before consumer and social justice groups began calling for boycotts of Limbaugh's sponsors, along with radio stations carrying his program. Large news bureaus began rebroadcasting his comments. And retailers and manufacturers started putting as much distance between them and talk radio as possible.

Then in February 2012, the man responsible for changing the face of AM radio called Georgetown University Law Student Sandra Fluke—who had given testimony before the US Congress regarding insurance coverage for contraception—a "slut" and "prostitute." The *San Francisco Chronicle* claimed three dozen advertisers canceled their advertising contracts and the few sponsors local stations were able to convince to advertise on his program wasted no time pulling their ads.

Today, every AM station in the country is in a fight for its life. Though the stations attract millions of listeners every day, advertisers are far and few. Unsold ad inventories are at historic highs, and ad rates at record lows. Many stations have resorted to selling air time to infomercial companies, as well as sketchy "pay-per-inquiry" advertisers just to keep the lights on. And it does not look as if the situation will turn around any time soon. Recently, the vice president of marketing of one of the largest packaged-food companies in the world told me, "Talk radio is on our 'no-buy' list. We don't need the trouble. There are plenty of other places we can spend our money." Another CEO echoed a similar sentiment: "We won't even buy advertising on *stations* that carry Limbaugh. We don't care if the advertising is on a different program than his."

Using talk radio as an example of an industry that has failed to adapt, let's look at the first seven principles at work:

Principle One. Failure is more common than success. In radio and television, many more programs fail than live to see a second season. Given this high rate of failure, was it wise for radio stations to lock themselves into expensive, long-term contracts with so many lookalike talk shows at one time? Bet their stations on content which appeals to only half of the consumers a sponsor wants to reach? Half a station's listening audience?

Principle Two. The greater the magnitude, speed, and complexity of change, the higher the rate of failure. The transformation of AM radio from music to talk happened in a short span of time. Today, you would be hard pressed to find one or two music stations on your AM band. Once you throw in the effect of deregulation, the advent of the Internet, and the spread of mobile devices on which you can now watch an entire movie into the mix, it's easy to understand why it has been difficult for AM radio to adapt.

Principle Three. Any drive toward singularity is a drive toward extinction. As advertisers grew fearful of Limbaugh's controversial remarks, they grew fearful of AM radio. The entire medium began to look risky. With no neutral, safer programming for advertisers to choose from, big advertisers put talk radio on their "no-buy lists" and stations kissed their largest source of income goodbye.

That said, a recent study reveals there may still be hope for talk radio. Radio stations offering a *mix* of programming—finance, science, automobile, and nonpartisan shows in combination with local news coverage appear to be doing better than their counterparts who run the one type of programming day after day. In other words, those practicing diversification are faring better than those who have opted for singularity.

Principle Four. Success warrants imitation. If radio stations had looked for a successful model outside of their industry, they would not have had to look any farther than A&E Networks' History Channel. Once a sleepy cable channel offering strictly *National Geographic*-like educational programming, the History Channel underwent a serious makeover five years ago—a makeover aimed at not only attracting more viewers but also big advertisers. Today, with programming like *Vikings*, *The Bible* and *Houdini* miniseries, *American Pickers*, *Ice Road Truckers*, and *Pawn Stars*, the network has expanded the definition of *history* to encompass a much wider spectrum. It is precisely this diversification which has propelled the History Channel to become one of the top cable television stations in the country.

Principle Five. The size of an environment determines growth. Though the radio market is larger than television or print, there is a limit to how many competitors any market can support at one time. With 15,432 radio stations in the United States competing for the same advertisers, it is likely AM radio stations will see great consolidation and attrition in the near future.

Principle Six. Nature's defenses: quick, cunning, and camouflage. The AM radio market adopted none of nature's time-honored defenses, despite being under attack from all sides. They failed to defend their territory from Internet and satellite encroachment by demanding that talk shows be exclusively broadcast locally and live. They failed to build customer loyalty programs that would have slowed the rate at which advertisers jumped ship. They failed to protect a business model on which their very livelihood depended.

Principle Seven. Nature's offense: strike first, strike hard. Make no mistake, radio stations could have fought back. Once they realized audience size had no relationship to monetizing programs like Rush Limbaugh's, they should have renegotiated their contracts, forcing controversial hosts to make revenue guarantees—guarantees such as buying back ads if a station loses an advertiser because of something a host said. At the very least, this would have mitigated the financial loss stations suffered after Limbaugh's controversial comments about Sandra Fluke and others.

And finally, we come to *Principle Eight: Incremental adjustments pose less risk.*

Had AM radio's transition been gradual and abided by the principles which govern successful adaptation, AM stations might have struck a balance between programming that has wider audience and sponsor appeal—similar to what the History Channel did. But the *Rush* to adapt all at once (can't help the pun), combined with a lack of foresight, put AM radio at the top of the endangered species list.

Principle Nine
Pattern recognition is vital to survival.

In nature, the ability to quickly identify patterns is a valuable advantage. Predators who correctly learn the habits of their prey are more successful than those who rely on chance encounters. Likewise, those who can detect the early signs of a threat greatly reduce their chances of becoming food.

A little over a year ago, biologists made an interesting discovery regarding insects and celestial patterns. They discovered the dung beetle (also called the scarab) relies on the strong light cast by the Milky Way to roll its ball of dung in a perfectly straight line to a safe place where it can be consumed. This navigation system prevents the beetle from inadvertently rolling the dung in a circle back toward the dung heap where other beetles would have an opportunity to steal it. It is remarkable that a tiny beetle depends on a pattern of lights twenty-six thousand light-years away for its daily survival.

Who knew?

In 2015, chairman of the World Conference on Disaster Management Paul Kovacs described one of the most remarkable uses of pattern recognition and foresight that I have encountered since beginning my research for the book—one that cemented my belief that the future will soon be known and manipulated to produce a better outcome. According to Kovacs, insurance companies have become so good at forecasting disasters, they are now using that foreknowledge to *preempt future claims*. In one instance an insurance carrier made telephone calls to customers who lived in an area where a large hail storm was imminent. The company encouraged their clients to put their automobiles in the garage and find shelter for other items that could be damaged before the storm hit. If, after receiving the call, the client acted pre-emptively, incurred no damage, and filed no claim, their insurance premium was reduced. The customers' willingness to circumvent the need for a claim lowered their risk profile. It was a win-win.

In this way, predictive analytics and pattern recognition are changing the way the insurance industry views itself. Analytics enable today's carriers to warn customers about an increase in home burglaries in their area so they can take additional security measures; the trajectory of a fire so homeowners can remove valuables; hazardous road conditions so they take alternative routes. The insurance industry is becoming more and more forward-looking, pro-active, and preemptive.

And if that isn't an example of using data to predapt, I don't know what is.

Principle Ten
Compensatory behaviors mitigate shortfalls.

The day I admitted I was always running five minutes late was the day I set all of the watches and clocks I own ahead five minutes. I have been doing this for so long now that when I glance at the clock, it never occurs to me I have five extra minutes. I take the clock at "face" value.

I lean on other compensatory behaviors to mitigate my shortcomings as well. For example, sometimes I know in advance that I am going to forget something. I am deeply troubled by the fact that my mind is a good error predictor— it knows in advance, and with great certainty, that I am going to forget, though I haven't actually forgotten anything yet. Why doesn't my mind simply correct for the error by storing a reminder somewhere where there's no danger of forgetting?

So now, when I am sure I am going to forget to take some-thing with me, I put the item in front of the door so I will trip over it before I leave it behind. Or I write a note and stick in on the front door.

I have this same problem when I go grocery shopping. With so many products and isles and displays and offers, I frequently leave without getting what I went there for. So I keep a list in the kitchen and religiously write down the things I need. But sometimes I have to put a note on the front door to remind me to take the list...

Not all unproductive behaviors or urges can be changed. We often set unrealistic expectations. It's the reason New Year's resolutions, diets, savings programs, and rehab clinics have become fodder for late-night comedians. And also why people who insist on changing others live in a constant state of frustration.

But here's how I see it: We may not be able to change every behavior or circumstance, but very often we *can* institute workarounds.

In business, compensatory measures are common. An employee may have tendencies to be territorial, to shy away from accountability, to be unreliable. When any of these or other proclivities get in the way, workarounds have to be instituted. Whether it's instituting new signoff procedures to rein in spending, random drug testing, or implementing checklists, successful companies acknowledge the role human nature plays in the workplace and take steps to mitigate undesirable attributes.

In the natural world, organisms also develop workarounds. It is common for dogs who are going deaf to bark much more. They are warding off potential danger they can no longer hear. Some birds have been observed delaying nesting during times of severe drought. And hibernation is a compensatory behavior for warm-blooded animals that do not migrate.

When a creature or organization is resilient—when they are adaptable—they are certain to discover a myriad of ways to adjust to changing conditions.

Principle Eleven
Critical mass is a prerequisite for change.

The relationship between critical mass and survival in nature is well understood. When the number of tigers remaining on Earth reached 3,200, we sprung into action, creating laws to protect the species, initiating breeding programs, and educating the public. The same goes for mountain gorillas, condors, legless lizards, and other species teetering on the edge of extinction.

When it comes to initiating social change, critical mass is too often overlooked. Nonprofits and organizations are the worst offenders, and this is the reason their impact is generally anemic. They fail to understand that their top priority must be recruitment. No recruitment, no critical mass. No critical mass, no change. It sounds like a simple idea, but when you see how quickly advocates become arrogant and combative and do everything to alienate those who don't agree with their position or methods, there is only one conclusion you can draw: they are not nearly as interested in succeeding as they are servicing their passions.

Yet history shows time and time again that nothing happens without critical mass. You cannot even get a second season of a television show picked up without it. Consider this: it wasn't until Dan Rather brought images of the Vietnam War into the living rooms of middle America that thousands of protestors hit the streets, bringing an end to US involvement in Vietnam. Critical mass was responsible for bringing Prohibition to an end; for Civil Rights; for prompting our daily recycling of paper, cans, and bottles; and recently, for the rise of the Arab Spring. It was the power behind the

American revolution, disco, and the spread of social media. In each case mass engagement was key.

Not long ago, the importance of mass acceptance was driven home in E. O. Wilson's landmark book, *The Creation*. Written in the form of a letter to a Southern minister, the greatest evolutionary biologist in the world reached across the great divide between science and religion, inviting both sides to work together for the good of the planet. While world leaders focus on the dangers of climate change, the loss of animal and plant habitat and related extinction poses a far more urgent problem. Scientists estimate we are losing species at a rate that is one thousand to ten thousand times greater than normal (baseline)—a dangerous rate of extinction the plant has not experienced for sixty-five million years. Against this backdrop, Wilson wrote:

> Pastor, we need your help. The Creation— living Nature—is in deep trouble. Scientists estimate that if habitat conversion and other destructive human activities continue at their present rates, half the species and animals on Earth could be either gone or at least fated for early extinction...
>
> Surely we can agree that each species, however inconspicuous and humble it may seem to us at this moment, is a masterpiece of biology, and well worth saving...
>
> You may well ask at this point, Why me? Because religion and science are the two most powerful forces in the world today... If religion and science could be united on the

common ground of biological conservation,
the problem would soon be solved. If there
is any moral precept shared by people of all
beliefs, it is that we owe ourselves and future
generations a beautiful, rich, and healthful
environment.

Wilson then posed a pivotal question: Is it possible that
the Garden of Eden was what existed before humanity came
along? And if so, *what can religion and science do together to
restore Eden? To restore the Earth to its natural condition?*

Now you may be wondering why an esteemed scientist like
Wilson felt compelled to make a religious case for biodiversity. I admit I wondered that myself.

So one day I asked.

Wilson explained, "Any movement which disenfranchises
people of faith can't succeed." According to the naturalist, this
was simply a matter of simple deduction. The latest Gallup
poll shows that over 90 percent of US citizens believe in God.
So how could a social movement that *excludes 90 percent of the
population* succeed? On the other hand, imagine the odds if
you compel that 90 percent—or some portion of it—to throw
down with you!

While other academics were busy looking down their noses
at believers, Wilson got busy finding a common purpose. He
asked religious leaders to set aside the question of whether
we descended from Adam and Eve or single-celled organisms
to focus on a more urgent question: Do worshippers have a
moral obligation to protect God's creation? And with that,
he earned the respect and help of the 90 percent.

In nature, there is power in numbers. Numbers mean survival. Similarly, organizations and movements that build critical mass greatly increase their opportunities for success.

Principle Twelve
Fortune favors the prepared mind.

The last principle of adaptation is about the role foresight and preemption play in adaptation. It would be difficult to name two greater advantages than the ability to foresee and shape future events. Yet we know very little about the evolution of human foresight—and even less about foresight in animals. Clearly my dog has the ability to project ahead when he sees the black suitcase come out of the closet, and he knows the beach or park aren't far away when I grab the leash off the laundry room wall. But so much of animal behavior is driven by instinct and necessity, it's difficult to know what their capacity for imagining future scenarios really is.

Anyone who has watched a tiger stalk its prey knows there is some form of premeditation which takes place before a strike. In the Northwest, a deer's coat turns a noticeably darker color prior to an abnormally cold winter. Sharks have been observed moving to deeper waters just before a hurricane strikes. And there are many reports of animals escaping to safety before an earthquake and tsunami occur. It is clear these animals are responding to data of some type, though we do not presently understand what that is. Theories range from an animal's heightened sensitivity to changes in atmospheric and water pressure, to their ability to detect infrasonic pulses. Michelle Heupel, a scientist at the Mote Marine Laboratory, explains their fore-action in this way: "I think these animals are more attuned to their environment than we give them credit for." And that sounds about right. They have data we do not.

At least not yet.

But what humans may lack in instincts, we more than make up for in technology and knowledge.

In 2013, I had an opportunity to interview William Scott, test pilot and former officer with the National Security Agency (NSA) about a little-known threat called pyro-terrorism.

In 2011, the Navy Seal Six raided Osama bin Laden's lair in Pakistan. In addition to capturing bin Laden, they made a disturbing discovery. There, in plain sight, lay detailed Al Qaeda plans to set US national parks and wild lands in Montana, Wyoming, Utah, and Colorado on fire.

An examination of the recovered documents revealed that bin Laden had come to the conclusion that Al Qaeda's attacks were too complicated, expensive, and difficult for the average person to carry out. They needed an easier way to do damage—one that didn't require getting a pilot's license, learning bomb-making skills, becoming an expert at disguises and hiding undercover, and so on. The ideal weapon was one which any person of any skill level could use at a moment's notice.

Arson turned out to be that weapon.

Arson was the best way to deliver a blow to the US economy for a number of reasons: A) parks and wild lands are unprotected and easily accessible, B) matches (and firebomb-making materials) are available anywhere, C) any person can strike a match, D) a penny match would produce hundreds of millions of dollars of damage, and E) drought conditions and US laws prohibiting the thinning of trees and removal of dead vegetation had combined to create ideal fire conditions. So Al Qaeda hatched a plan to burn, burn, burn.

Around the same time Al Qaeda was educating their followers on the virtues of pyro-terrorism, fire departments in the United States began reporting a rise in human-initiated fires. Then the director of Russia's Federal Security Bureau Aleksandr Bortnikov announced Al Qaeda had been complicit in a number of coordinated forest fires in Europe as part of their "terrorism strategy of a thousand cuts." This was all Scott needed to hear.

There was no reason to wait until every state was burning.

But averting arson is not as easy as it sounds. You can't outlaw matches. Or monitor every hiker who enters a national park. How do you stop a hundred terrorists from igniting fires in remote forests at the exact same time?

Two years after discovering Al Qaeda's plans, Scott began urging lawmakers to change the oversight of the National Park Service. Today, these parks are considered a land management issue. So, they report to the US Department of the Interior—an agency that is untrained to deal with national security. Scott urged Congress to move the responsibility for protecting parks to Homeland Security.

Next, Scott put forward the idea of using planes, drones, and satellite technology to watch for "hot spots" inside parks and other areas known to be targeted for arson.

This would allow fires to be identified and put out when they were small. The drones would also be able to photograph and track the bad actors. Scott made the point that if terrorist-instigated fires were quickly extinguished and succeeded in doing little to no damage, this would discourage other pyro-terrorists. Pavlov proved that animals abandon behaviors that produce negative or no results.

Lastly, Scott urged the US government to build a "robust large air tanker fleet of firefighting aircraft" that could be deployed within minutes of a reported outbreak. The equipment currently used to fight forest fires is outdated, unreliable, and very dangerous. In two incidences, the wings fell off off planes en route to California and Colorado. To prevent pyro-terrorism from succeeding, a new, emergency-ready fleet operated by a specially trained National Guard-like force was needed.

Six years after discovering Al Qaeda's plans to set the United States on fire, little has been done. Though Scott constructed steps to eliminate or minimize the danger, we hesitate to act. But this is not largely different from how we have responded to new information about other dangers we see coming with greater clarity than ever before. We have the evidence. The algorithms, models, and forecasts. The means to prevent what has yet to occur. Then why don't we? What stands in the way of leveraging the principles of adaptation to our advantage? In the way of using foreknowledge to predapt?

"It is impossible to overlook the extent
to which civilization is built upon
a renunciation of instinct."

- Sigmund Freud

CHAPTER SIX

The Invisible Tether

A group of research assistants charged with teaching an ape to use sign language decided to conduct an unauthorized experiment.

Every day at 3:00 p.m., the assistants were instructed to bring a snack to the ape. They entered the cage and presented a tray of food choices (green beans, cucumbers, bananas, etc.). The ape indicated its food selection using sign language—"green slices" for cucumber, "red circle" for tomato, and so on. Though the ape had a limited vocabulary, its intentions were easily understood.

The researchers spent hours every day playing, feeding, bathing, and teaching language to the primate, so they developed a close bond. But they wanted to put that bond to a test. They wanted to know if—given the opportunity—the ape they had grown so fond of would deceive them.

One day, instead of the regularly scheduled time for a snack, a researcher entered the cage an hour earlier with their normal tray of options. Surprised and delighted, the ape looked over the snacks, made a selection, and quickly consumed the food. Then the researcher exited the cage. An hour later—at the normal snack time—a different researcher—again in a white lab coat with a tray of snacks—entered the cage. But before presenting the tray to the ape, the researcher explained that they had heard someone had already given the ape their snack. The researcher asked the ape if this was true:

"Did you eat your snack today?"

The ape gave a curious look, shook its head, and quickly signed, "No. No snack."

Then the researcher offered the primate a second chance—an opportunity for redemption. "Are you sure you didn't eat your snack?" they signed.

It was at this moment the ape's expression became sad and hurt—the animal expressed the same emotions any person unjustly accused would.

"Snack? No. No eat snack. Me no eat snack," pleaded the ape.

So the assistant presented the tray to the primate, who chose a banana and gleefully consumed the bonus bounty.

But as the researcher left the cage that day, something had changed. The ape they had come to adore had looked them straight in the eyes and lied.

Free will is not always free. Many of our allegedly free decisions are driven by prehistoric instincts and unconscious motivations. Does an ape have a "choice" to tell the truth if telling the truth denies it food? Not where their DNA is concerned. Food is necessary for survival. Honesty is not.

That's a tough pill to swallow. We want to believe we have the power to choose and act in all matters. We can alter our behavior any way we want—use foresight to adapt *at will*.

But by how much?

For over a century we have been struggling to come to terms with the fact that *humans are born with predispositions*. We've struggled with just how much inherited programming can be overridden—as well as how much organizations, governments, and social systems can alter the DNA of their origins. Can Dole transform itself into an ER? Can the US Department of the Interior act more like Homeland Security? Can a child sociopath become a trustworthy adult? Yemen a democracy?

In theory they can...

That said, more often than not *biological obstacles* stand in the way. We find ourselves jerking, pulling, and struggling against an invisible tether that binds us to our past. Behaviors that were once foundational to the success of our prehistoric predecessors now hinder progress.

Take a simple reflex like fight or flight.

On the one hand, millions of years of evolution have produced technology that permit us to accurately preview and alter future events to suit us. On the other hand, we have evolved *no physiological response to long-term danger*. A fire, a snake, the sound of footsteps behind us at night—these things cause chemicals to flood our bodies and compel us to take action. Our amygdala triggers a reaction in the hypothalamus which signals the pituitary glands to release adrenocorticotropic hormones (ACTH). Then, at that exact moment, our adrenal glands produce epinephrine. Together ACTH and epinephrine create a surge of energy. Within milliseconds our blood pressure and sugar soar, our immune system kicks into hyperdrive, blood is diverted to our muscles, our pupils open wide, we perspire to cool down our body temperature, fat cells release stored energy, and our muscles tense.

Boom! We're ready.

But sit anyone down and explain the dangers of pyro-terrorism, the loss of biodiversity, deficit spending, and nuclear proliferation and we experience no change in our physiology. Though these threats are far more worrisome than a snake, they're not imminent. There's no urgency to act. So these and other long-term threats become the subject of endless meetings, analysis, presentations, planning, and debate. Though we have an uneasy feeling the clock is running out, we can't seem to get it together until the building is on fire. All the foresight in the world is for naught.

Population growth is a good example of just how difficult it is to tame evolutionary obstacles which stand in the way of progress. If we know our numbers are growing, and we have computer models that show just how many human beings the Earth can reasonably support, then it follows we would use that information to circumvent catastrophe. No need to wait until starvation, disease, and violence are upon us. We can use the foreknowledge to predapt.

Growing, Growing, Gone

In 2016, news bureaus reported that our numbers would approach 7.5 billion within one year. They noted it took 150,000 years to create the first billion humans. The second billion around 123 years. And the most recent billion? Only twelve years. If population growth were a stock on Wall Street, investors would be lining up.

Though news bureaus have been reporting that the rate at which we are populating has recently declined in several industrialized nations, make no mistake about what the decline means. In some locations growth has slowed—but our overall numbers continue to rise. The fact is, scientists estimate there is a greater than 80 percent probability we will hit between 9.6 to 12.3 billion people by 2100.

Predictive analytics now leave little doubt as to the resources required to sustain the projected increase. Joel E. Cohen, the author of *How Many People Can the Earth Support?* reveals that today, it takes approximately 10 hectares of land to produce enough food, water, medicines, and shelter for the average American. To bring the rest of the world to a similar standard of living, we need close to 70 billion hectares of land. But here's the catch: the total combined land mass of Earth only adds up to roughly 15 billion hectares (never mind useable land). So we would need four additional Earths to bring everyone up to the standard of living enjoyed by one country today.

Four additional Earths?

We don't need a Big Data system to foresee the consequences of continuing down this path. Either the standard of living for all people will decline across the board: the gap will widen between the haves and have-nots; a pandemic virus, war, meteor strike, volcanic eruption, water and food shortage, or some other disaster will quickly diminish our numbers; or improvements in farming, energy, potable water, etc., will buy us time. We can anticipate that one, or some combination, of these outcomes will occur. That's because there's no fighting the *Fifth Principle of Adaptation: the size of an environment determines growth.*

Fortunately, to this point, human ambition has saved the day.

Since the early twentieth century, mechanization, pesticides, and synthetic nitrogen and phosphates have significantly increased crop yields. Higher yields also meant cheap surplus for livestock, so cattle and pig ranches, chicken farms, and dairies could expand. And the transition to high-performing, disease-resistant varieties of rice, wheat, and corn (60 percent of a modern diet) has had an impact on global food production. Thanks to these and other innovations, agricultural productivity in the United States is three to four times greater than it was in 1948. And that's just one nation.

But recently agriculture has hit a ceiling.

Global yields for rice—a main food staple—have not improved since 1966. The same goes for corn—no change for thirty-five years. What's more, the nutritional value of food is on a sharp decline as soils become depleted and modern additives fail to compensate for the shortfall. In 1950, broccoli had 130 mg of calcium. Today, it has less than 48 mg. So, in addition to more mouths to feed, we also have to eat more to get the same nutrition.

But food is only one of many challenges posed by population growth. It's difficult to turn a blind eye to other problems this produces. Increased unemployment? If the number of people who need work doubles every few decades, it follows that unemployment will go up. Deteriorating infrastructure? If twice the number of people are driving on roads and bridges and using twice the electricity and water, infrastructure is going to wear out much faster. And how about problems like climate change? Is there any doubt the jump from seven to twelve billion people means more human activity? More carbon emissions, deforestation, heat generation, extinctions? Even a rise in the number of homeless is attributable to population growth. Today, more than five million people sleep covered with no more than a newspaper on the streets of Mumbai, India. And similar conditions exist in Mexico City, São Paulo, Shanghai, Karachi, and Dhaka. The latest estimates put global homelessness at two hundred million people—and the number is climbing faster than expected as millions of refugees flee Syria, Afghanistan, and other war-torn countries. Last year the United Nations reported the number of displaced persons hit a new high, exceeding the entire population of Great Britain: over sixty-five million men, women, and children are on the move.

The obvious solution to these and other social problems is to stop reproducing in numbers that cannot be reasonably accommodated.

Here again, science, technology, and human ingenuity offer remedy—bowing to the *Tenth Principle of Adaptation: compensatory behaviors mitigate shortfalls.* That which we cannot change can be mitigated.

Since 1855, when the first rubber condom was produced— and the early '60s when the first birth control pills were introduced—we've had the means to avert overpopulation. If we include methods such as abstinence, then we've had the ability to control birthrates since before latex and pharmaceuticals arrived on the scene. So what's the problem? Why have all our efforts—from sex education, free contraception, sterilization, tax incentives, and using the full force of the law to prohibit sex among young people—barely made a dent?

The answer is evolutionary mandates. The more resistant a problem is to cure, the more likely it is the resistance can be traced to the vestiges of prehistoric instincts on which our survival once depended.

We are far more comfortable blaming our failure to stop overpopulation on a lack of leadership, weakness of character, poor parenting, a lack of education, or the expense or unavailability of birth control. While these factors can play a role, we ignore the most obvious reason our countermeasures have failed: *next to survival, the drive to procreate is the strongest drive living organisms possess.* When we attempt to suppress something as powerful as propagation, you better believe we're in for a fight. We're taking on Mother Nature herself!

It's no wonder we find ourselves at war with the part of our brain that knows better. Do we expect reason and foresight to intervene at that magic moment when our pleasure centers, our hormones, and every nerve ending in our body is screaming "Go! Go! Go!" What other explanation do we have for a leader putting the highest office in the world at risk for a dalliance with an intern? The head of the World Bank assaulting a maid in his hotel room? A television evangelist turning his back on the gospel?

Paleolithic Propensities

How much *do* inherited predispositions inform our decisions in modern times? How strong is our genetic tether? Stronger than we are willing to acknowledge.

Last year, I was invited to speak at a conference which featured a star-studded list of celebrities from around the world: Nobel prize winners, heads of state, corporate innovators, media personalities, and so on. Though I was honored to be included on the roster, I was keenly aware that I was outmatched in every way. These were people who were more accomplished and better known, so I hoped to keep up and make a respectable showing. The conference was organized so multiple presentations were offered at the same time. Attendees selected the topics and speakers they wanted to hear from an à *la carte* menu, so if two speeches were scheduled at the same time, an attendee would have to choose one.

Since the other presenters had a large fan base, I prepared myself for a small turnout at my speech. But when I arrived at my designated room I was surprised to see it was overflowing with participants. People were lined up against the walls, sitting in the isles—some were using the tables in the back of the room as a bench to sit on. The room was animated and noisy with anticipation.

So I doubled back to make sure I had the right room.

Thankfully, my face was plastered on a flyer at the entrance, so, I grabbed one and made my way to the stage. As I took a seat and waited for the host to do the introduction, I grew more curious. More people were pouring into the room. The hotel staff were grabbing chairs off a cart and setting them up as fast as they could move. The excitement was building.

Then—just as my host stood up to welcome the audience—I glanced down at the flyer in my hands. The title of my speech was "The Future of the Human Organism." But there, at the top of the page in boldfaced type—impossible to miss—was the typo to end all typos: "The Future of the Human *Orgasm*."

Case solved.

Just when you're certain the world is waking up to the role evolution plays in the foibles of modern man—BAM! Human nature comes raining down like a ton of bricks.

Since there was no time to change my speech I stepped to the podium, thanked my host, looked across the room, and asked, "How many of you are here to hear about the future of the human *orgasm*?" All hands shot up in the air. Then, with a smile, I held the flyer up and said, "And—just out of curiosity—how many came to hear about the future of the human *organism*—the title of the speech I was *supposed* to give tonight?" The room broke out in laughter. "Okay then," I said, "The *orgasms* have it. Fortunately, it's the same speech. The future of the organism depends on the future of the orgasm—doesn't change a thing..."

Whether we're picking an afternoon lecture, or battling a behemoth problem like population control, often our behaviors are influenced by inherited programming. It may be convenient to pretend our bodies have cleansed themselves of prehistoric inclinations, but this simply isn't factual. In truth, the more we deny our genetic predispositions, the further we drift from solving problems which have dogged humankind for centuries.

And nowhere have we ignored the evolutionary origins of our consternation more than when it comes to ongoing instability in the Middle East.

We've blamed religion, dogma, and tradition. We've blamed political corruption. We've blamed culture and education. We've blamed oil and big business. But no one has considered the persistence of prehistoric imperatives such as territoriality, reproductive urges, troop behavior, etc. as the *root cause* of turmoil and unrest.

Well, why not? Surely a conflict that has persisted for thousands of years must have some biological basis.

The Arab Spring

At the end of 2010, the suicide of a twenty-six-year-old Tunisian street vendor ignited the biggest revolution of the twenty-first century.

That said, the story of Mohamed Bouazizi is not the story of a hero. It is not the story of a revolutionary, martyr, or individual blessed with exceptional foresight or talent. Not the story of intention or achievement.

Mohamed Bouazizi—the middle son of a family of six—lived on the outskirts of the small town of Sidi Bouzid in Tunisia. When he was three years old, his father died of a sudden heart attack. His mother considered herself fortunate to have quickly remarried her husband's brother—a blessing that saved the family from destitution. But the marriage was not enough to avert poverty. By the time Bouazizi turned ten, he and his siblings were forced to work the streets of Tunisia to keep the family alive. Bouazizi didn't mind. He was well liked, and over time, was able to parlay his knack for salesmanship to selling fruit from his own wheelbarrow. This allowed Bouazizi to earn a livable wage of about $140 a month.

According to his friends, Bouazizi was happy to have any job. At the time, Tunisia's unemployment rate was over 30 percent and the country's GDP had dropped 50 percent in twenty-four months. Even jobs that paid below subsistence were difficult to find. Making matters worse, there was no indication the Tunisian government had plans for improving conditions. The situation for young men in Tunisia was turning more desperate by the minute.

Though Bouazizi lived in an apartment no bigger than a closet in one of the worst slums of the city, he was better off than others who could not pay rent and had no safe place to rest their heads. According to other street vendors, Bouazizi was hardworking, diligent, and cheerful, and known to offer his unsold fruit to the homeless on whom he took pity. Similar to other vendors who were also barely getting by, when Bouazizi was not selling fruit he was hustling for other work—any work that might ensure his survival.

Law enforcement authorities were also feeling the effects of inflation. As the economy grew worse they began harassing street vendors for money. They claimed the vendors needed special permits to sell their wares on the street. But authorities were willing to look the other way if the merchants paid them cash.

Sadly, Bouazizi couldn't afford to pay the bribe. He was barely able to pay rent and whatever was left over was needed to procure more fruit to sell. But this explanation did little to stop the police from terrorizing Bouazizi every morning and afternoon. And as other vendors began to acquiesce, the threats against Bouazizi grew more pronounced.

Recognizing his livelihood was in jeopardy, Bouazizi came up with a plan. He borrowed money to purchase more fruit— optimistic that he could sell the additional fruit to pay the officials. But now, in addition to owing a bribe, Bouazizi took on the high cost of street interest.

The pot was beginning to boil.

Bouazizi worked day and night to sell enough fruit to pay the lenders and authorities, who, according to witnesses, began targeting Bouazizi—pushing his wheelbarrow to the ground, shoving him, kicking him and screaming threats. Bouazizi begged for patience, but the officials showed no sympathy. In fact, the more he tried to reason with them, the more intent they became on using him as an example.

Then one day, the authorities confiscated Bouazizi's wheelbarrow of fruit. It was their final attempt to extort payment.

Broke, exhausted, and humiliated, with street lenders now on his heels and no means to survive, Bouazizi made repeated visits to the police station that was holding his wheelbarrow, scales, and merchandise ransom. He pleaded with them to allow him to return to work.

On December 17, Mohamed Bouazizi made a final plea to the governor.

When he was turned away, he walked from the governor's office to a nearby gas station and used the remaining coins in his pocket to purchase a can of gasoline. Then, amid the mid-afternoon traffic, he doused himself and lit his body on fire.

That evening, as Bouazizi lay dying in a nearby hospital, his mother led a peaceful protest on behalf of her son. And from here, Bouazizi's story spread quickly. Soon street vendors and their families joined with young men who could not find work and shopkeepers who could no longer afford bribes. They poured into the alleys and streets and the squares of Sidi Bouzid. The news of Bouazizi's suicide found its way to other towns where similar oppression had become a part of daily strife. And suddenly, an entire nation was on the move. Ordinary working men and women began crowding into town squares, cutting off roads and commerce, demanding reform. The *Eleventh Principle of Adaptation took hold: critical mass is a prerequisite for change.*

And this is how the Arab Spring was born.

The self-immolation of Mohamed Bouazizi became no less a powerful symbol as Rosa Parks' decision to sit at the front of the bus, Ghandi's nonviolent protests in India, or Joan of Arc's battle cry. When society finds its voice in a heroic act of defiance, it cannot be silenced.

But it is here that our story goes awry.

The protestors who filled the streets of Tunisia, Egypt, Yemen, Libya, and other countries attributed high unemployment, poor educational opportunities, runaway inflation, human rights violations, and deteriorating infrastructure to corrupt governments. So, from their perspective, the solution was simple: replace a repressive regime with a representative democracy. Do this and conditions would improve.

No one seemed concerned with the *Second Principle of Adaptation: the greater the magnitude, speed, and complexity of change, the higher the rate of failure.*

By January 2011 the revolution had succeeded. The president of Tunisia, Zine el-Abidine Ben Ali, fled to Saudi Arabia, and a month later, under similar circumstances, the president of Egypt, Hosni Mubarak, stepped down. Libyan leader Muammar Gaddafi was overthrown, the leaders of Bahrain, Syria, Kuwait, Lebanon, Oman, Jordon, Morocco, Algeria, and Jordan announced constitutional reforms, and the president of Yemen, Ali Abdullah Saleh, handed the reigns to a successor. New government reforms and programs aimed at alleviating human suffering quickly fell into place, and the air was filled with the intoxicating perfume of hope.

But four years later, nothing had changed for the person on the street.

In Egypt, the same protestors that once gathered in Tahrir Square to demand the ouster of Mubarak gathered again to demand the military—who was acting as a transitional government—turn the reigns over to civilian control. All the while, unemployment continued to rise, schools and infrastructure continued to deteriorate, and poverty continued to spread.

Despite the courage of working men and women who swore their allegiance to a new democracy, the Arab Spring did nothing to alleviate misery. Soon the tender membrane of optimism was pierced by the reality that it didn't really matter whether Mubarak or the Muslim Brotherhood or any other leader was in charge—the source of Egypt's problems had much more to do with biological imperatives than governance or power.

Over forty years ago, scientists saw the Arab Spring coming. But it was not dictatorships or democracies that concerned them. It was the rate at which the population in the Middle East was exploding. They were concerned about the instability that doubling the number of citizens every one or two decades would lead to.

Using Egypt as an example, in 1981, when Hosni Mubarak became president, the country's population was south of 40 million. Today, the population of Egypt is approaching 95 million. This means that in order to prevent Egypt's standard of living from sliding backward, Mubarak would have had to generate twice as many jobs, twice the number of roads, hospitals, sewage treatment plants, schools, utility plants, and homes, and double the capacity of every government agency, all in three decades. And accomplish all this on the back of a dwindling tax base.

No ruler in modern history has achieved such a feat.

Adding to Mubarak's challenge was the fact that less than 3 percent of Egypt's land is suitable for growing crops or raising livestock. A thin strip along the Nile River represents the sum total of arable land on which the population must subsist. Every year the strip becomes less productive as minerals and nutrients become depleted. So, while the population was growing at an unstoppable rate, the country's ability to feed and service the population was diminishing.

A quick review of the conditions of other Arab Spring nations reveals a similar pattern: population growth faster than governments could adapt. There was no opportunity to evoke the *Eighth Principle of Adaptation: incremental adjustments pose less risk.*

Which begs the question: when did problems *caused* by the drive to reproduce become a simple matter of ousting a dictator? Why did protestors—many of whom put their lives on the line—believe constitutional reforms would solve their problems? What made them think free elections and better leaders could produce twice the jobs, schools, roads, and hospitals? More arable land?

Six years after Hosni Mubarak peacefully handed the reigns of Egypt to a new government, the country's problems persist. That's because democracy is no antidote for scarcity. In truth, voting at the polls will do far less to fix Egypt's problems than condoms will. And with the United Nations forecasting the country's population to crest 115 million in a few years, Egypt's troubles will continue to magnify.

Taming the Beast

So, what's the answer? What can leaders in Egypt and other overpopulated nations do to predapt? To manipulate the future to ensure success?

Well, for openers, you can't fix a problem you don't admit. Given what we now know, there is only one way for Egypt's leadership to effect meaningful systemic change and do it quickly. They must come to terms with the fact that there are more Egyptians than the territory can sustain. Leaders must do the unthinkable: they must ask the same people who risked their lives for freedom to leave the country. Not just one or two. Millions of citizens must go elsewhere.

Think about it. If a quarter of Egypt's population could be convinced that leaving was the patriotic thing to do, this would produce *immediate* job openings, there would be a sufficient tax base to support those who need government aid, the country's infrastructure would be appropriate for the number of citizens, there would be sufficient housing and schools. The mass exodus would buy time for the economy to catch up, to stabilize—so that systems could be built to accommodate more citizens in the future.

While, at first blush, this may sound like a radical idea, the truth is it's not a new one. Though never officially sanctioned by the government, this strategy worked for Mexico—a nation that struggled with similar population growth, high unemployment, and corruption for decades. Estimates vary on how much money Mexicans living in the United States send to their families in Mexico, but experts' recent estimates suggest it is over $23 billion each year. This influx has been one of the largest, most effective economic stimulus plans Mexico has ever implemented. It has helped lift Mexico's GDP to $1 trillion in six years, allowing the country to boast a lower unemployment rate than the United States. What's more, key indicators for literacy, health care, and infant mortality are also all up in Mexico—offering further proof that exporting a nation's population is an effective short-term solution to runaway population growth.

But encouraging citizens to leave their native country is a temporary fix. It only curtails a symptom, because it does nothing to stop the programmed biological urges responsible for the growth. After all, there's no mystery as to *how* Egypt became the most populated country in the Middle East. We know *how*. And we also know that left unchecked, the population will double and triple again and the country will be back to square one. So in concert with mass relocation, the government must implement an aggressive contraception program—including free sterilization clinics. Economic carrots and sticks (tax incentives) must be used in conjunction with sex education and every other trick in the book.

If ten or twenty million Egyptians could be incentivized to move out of the country and seek jobs elsewhere, if birth rates were curbed, if borders could be secured to prevent foreigners from coming into the country to take advantage of jobs and services which result from the exodus, then it follows that the problems that drove citizens to protest in Tahrir Square would be remedied. Jobs would open up. Streets and schools too. There would be fewer mouths to feed, fewer patients in hospitals, fewer prisoners and homeless.

But it will never happen. Even if it makes all the sense in the world. What leader has the political courage to ask citizens to leave their country? Encourage contraception when religious tenets prohibit it? Act on behalf of the greater good at the expense of reelection, power, and legacy?

With no plan to address evolutionary obstacles, new governments and promises will come and go. But little will change in overpopulated countries for people like Bouazizi. Soon the Arab Spring will become the stuff of folklore—a subject for academics, a footnote in history, another example of man's failure to address the evolutionary tether that inhibits real progress.

Biobstacles Everywhere

Once we see our struggles from the vantage point of who and what humans are, predaptation is made easier. We can use foresight to act prophylactically and head problems off at the pass.

For example, using what we know about human behavior, we could have headed off the problem of texting and driving rather than waiting until the number of accidents was sufficient to act. Even to this day many of us are convinced we can type on a tiny keyboard while simultaneously controlling a two-ton vehicle at sixty miles an hour. We reject the simple fact that our eyes and attention cannot be focused in two places at once. In 2010, the US Department of Transportation reported that cell phones were involved in 1.6 million accidents and more than six thousand deaths. So what did we do? We made laws to stop people from doing what came naturally. We made it illegal, evoked stiff fines, installed cars with voice-activated phones, and today, made navigation systems inoperable when a car is in motion. If drivers won't accept their biological limitations, then legislatures and automakers will have to impose them.

But distracted driving is the tip of the iceberg—we've also come to believe we can live with a lot less sleep. But let's be clear here. Just because we lead busy lives and don't have time for sleep does not mean our bodies don't require it to function. In 2011, it took the fifth air traffic controller falling asleep on the job to prompt the Federal Aviation Administration to admit they had a problem with fatigue. When asked about the number of hours air traffic controllers were asked to keep their bodies in a high-stress, ready state, Philip Gehrman, director of the Behavioral Sleep Medicine Program at the University of Pennsylvania observed, "It would be nice if there were a greater appreciation that our bodies have a limit—we're not equally able to function at all hours of a twenty-four-hour day."

No kidding.

The difficulty we have connecting population growth to lack of opportunity, homelessness, climate change, and revolution—the difficulty we have connecting biological limitations with distracted driving, or air safety—all flow from the same faucet: our failure to account for who and what the human organism is at this point in time. Our physical and behavioral inheritance is rarely taken into consideration *before the fact*. So—by the time it is—we're in the thick of it, scratching our heads, wondering what went wrong, digging our way out as fast as we can...

One Banana, Two Banana

Our prehistoric inclinations not only influence our legal and social policies but also impact commerce in invisible ways. *Businesses that challenge genetically programmed behavior are far less likely to perform well than those that cater to natural proclivities.* By catering to our predisposition to use fashion to express our identity, Apple was able to overcome inferior performance when compared to their PC counterparts. Steve

Jobs was obsessed with industrial design because he knew the value of positioning the iPod, iPhone, and other products as fashion accessories. Similarly, pharmaceutical companies that began coating bitter-tasting medicines outperformed those that expected consumers to bite the bullet and ignore taste. UPS was the first to acknowledge the value humans place on foreknowledge by permitting consumers to track their parcels in real time. And in the last decade, Internet dating companies kicked the process of mate selection wide open. According to the Pew Research Center, 27 percent of adults between the ages of eighteen and twenty-four use online dating, and the number of users between the ages of fifty-five and sixty-four has doubled since 2013.

There is an evolutionary imperative behind every business success. And also one behind every failure, which Webvan, one of the richest start-ups in history, found out the hard way...

Webvan was founded in 1996 by Louis Borders—the entrepreneur known for launching the successful Borders and Waldenbooks stores. Borders had two years to study the impact Amazon was having on book distribution. Convinced the same model could be used for other products, Webvan grew out of a simple idea: deliver groceries to a customer's doorstep within thirty minutes of online ordering. Since marrying convenience with food has been successful since the beginning of humankind, a grocery home-delivery service was a no-brainer. And Benchmark Capital, Sequoia Capital, SoftBank Capital, Goldman Sachs, Yahoo!, and a long list of other prestigious venture firms agreed. Webvan couldn't miss. The company quickly became the best-backed start-up in the world, raising nearly $1 billion in cash at their initial public offering (IPO).

Then, five years after opening their doors, and eighteen months after their IPO, Webvan declared bankruptcy. It burned through all $1 billion with nothing to show but a trail of empty warehouses and vans across twenty-six cities. Webvan went from the richest dot-com to the biggest dot-com flop in history.

Analysts have offered many explanations for Webvan's extraordinary fall from grace. Some blamed the overexuberance of investors, which caused the company to become too rich too soon and recklessly squander their cash. Others cited the management team's lack of food experience. Still others claimed Webvan was ahead of its time and, had it not burned though all its money, it could have waited the market out.

While these may all have been contributing factors, it's hard to ignore the fact that Webvan was not the only grocery delivery company that failed to live up to the hype. PublixDirect closed. HomeGrocer was losing so much money that Webvan acquired it in a fire sale. (That should have been a clue.) British online grocery giant Ocado admitted that it had not been able to turn a profit in over ten years. Peapod, SimonDelivers, Gopher Grocery, Mr. Case, Grocery Gateway, FreshDirect, YourGrocer.com, and ShopRite were also struggling to keep their heads above water. After a decade of more failed attempts than successes, no one has been able to explain why buying groceries is not like buying books, clothes, and stocks.

But listen closely to one person's experience ordering groceries online:

> Well... the first time I tried ordering groceries
> [online], I ordered bananas. Under "quantity"
> I typed in the number ONE—thinking that

meant *one bunch*. And when the bag came,
there was one banana in the bag. So I realized
that you have to be more specific. Since I eat a
banana for breakfast every day, the next time
I ordered I typed in the number seven, for a
week's worth of bananas. But when I opened
the bag there were seven green bananas. They
needed time to ripen. So then I went online to
see if there was a place to say you wanted ripe
bananas you could actually eat. And about
this time, I realized there are things you can
order like toothpaste and canned soup but
there are others where it's just easier to just
go to the store and get what you want. So
why bother ordering any groceries online if
you have to go to the store anyway... it's like
shopping twice.

One customer I interviewed ordered a quart of almond
milk online. She received a quart of milk and a bag of almonds.
Another complained the expiration date on the tub of cottage
cheese meant she'd have to eat the entire tub that day. Some
consumers expressed disappointment in the selection—there
were not nearly as many options as appeared on store shelves.
And no one I spoke to thought online grocery shopping was
as good as going to the store.

Well, why not? Why is food different from other the other
things we order online?

Think back. In prehistoric times a bad food choice—such
as the inability to distinguish between a poisonous mush-
room and safe one, or rotten food from fresh, nutrient-rich

sustenance—often meant the difference between life and death. This is why we stay clear of apples with spots on them, deformed vegetables, and dirty produce. It's also why we're put off by unpleasant smells such as seafood that smells too "fishy." We are even programmed to avoid certain textures, such as a squishy persimmon or foods that are so hard they may endanger our teeth (which we not only needed to eat but also for protection).

Humans are very selective about what we ingest. And if you don't believe me, ask any grocer why the trash bin in the back is filled with produce that is perfectly fine to eat but doesn't look nice.

And there's a second reason online grocery shopping was destined to stall. All humans are born with a natural desire to hunt and gather—a behavior which has persisted for thousands of years. Stand in any grocery store and watch as shoppers pick up individual pieces of produce and carefully inspect them before adding them to their carts. Watch as they squeeze packages of toilet paper for softness, check labels for ingredients, scan the fifty types of bread offered. The truth is Mother Nature programmed humans to enjoy hunting and gathering—or what we call "shopping" today. *Shopping is in our DNA!*

Which explains why people still go to retail stores, farmer's markets, malls, and garage sales at a time when everything—even previously used items—can be purchased online.

That said, one day soon an Internet grocer will succeed where Webvan failed. They will study and leverage our genetic predispositions rather than ignore or alter them. Perhaps they will use a Big Data system or artificial intelligence to understand our individual preferences better. Or use virtual reality to walk through the imaginary isles and place virtual items into our virtual grocery cart. Perhaps facial recognition software will take note of our reactions to products and prices inside that virtual reality, then use this information to offer up special coupons and bundling opportunities. Or drones will deliver more than one choice of banana (green, ripe, and mixed bunches) from which we can select. Or 3-D printers will make it possible to manufacture most of what we eat at home and grocery stores will be relegated to raw ingredient suppliers. Tomorrow's innovators will find a way to design an online grocery shopping experience that is in tune with our natural predispositions. And rest assured, when they do, we will adapt lickety-split.

That's the beauty of *bioleverage*.

"When dealing with people,
remember you are not dealing
with creatures of logic, but with creatures
of emotion, creatures bristling with prejudice,
and motivated by pride and vanity."

- Dale Carnegie

CHAPTER SEVEN

Bioleverage

It was 7:00 a.m. and someone had to make the call: schedule the flight for another day or try to beat the storm in. As the crew stood in the hangar and watched the twenty-five-year-old US Air Mail pilot walk the dirt airstrip, they had a good idea what he would decide.

He'd need every inch of the runway to clear the power lines. And he'd have to put his faith in two engineers from the Ryan Aeronautical Company—fellows that outfitted the plane in less than sixty days and now didn't have the heart to look him in the eyes. Above all, he needed to stop thinking about the six who died attempting the same feat. Was he better than they were? As he turned back toward the hangar, he mumbled, "Just get her over the power lines... that's all... up and over."

A wiser man would have waited for better conditions. But he owed the State National Bank $15,000—and pilots with more experience were closing in—formidable challengers like Arctic explorer Admiral Richard Byrd and air racer Clarence Chamberlin.

He couldn't take the chance the $25,000 prize would be gone before he had a shot. So he ordered the men to remove the parachute, the sexton, and the radio to lighten the plane's load and then he took his seat behind the wheel of the small, fabric-covered monoplane dubbed The Spirit of St. Louis.

Thus began Charles Lindbergh's 1927 solo flight from New York to Paris. It took the pilot thirty hours to cross 3,600 miles. He escaped storm clouds at ten thousand feet, battled freezing ice, and flew through blinding fog for dispiriting stretches. In those days, pilots depended on the heavens to point the way, so a lack of visibility could send a plane hundreds of miles off course. Fuel, headwinds, and mechanical failures were also concerns. So by the time The Spirit of St. Louis touched ground at Le Bourget Airport on the morning of May 21 to 150,000 cheering spectators, there was no question the record and bounty were Lindbergh's to claim.

Whether it was the young pilot's fearlessness and good looks, or the entrepreneurial furor that swept over America in the 1920s, historians credit Lindbergh with igniting an international aviation boom. Overnight applications for pilot licenses tripled, and aviation companies found themselves flush with new investors. Within three years, the number of airline passengers jumped 3,000 percent.

But in the days following Lindbergh's historic achievement, the man behind modern air travel, financier and hotelier Raymond Orteig, was forgotten. The more time that passed, the more historians romanticized Lindberg's daring journey. Books were written and movies made, and his courage became the stuff of folklore. All the while, the $25,000 purse which raised The Spirit of St. Louis and her human cargo into the air grew evermore a footnote.

To his credit, Lindbergh never denied the fact that the cash reward motivated him to take to the air that fateful day. As he sunk deeper into debt, and more competitors began throwing their hat in the ring, the pressure mounted—until one morning the prize was more important than his parachute or sexton or an incoming storm.

What is it about rewards like the Orteig Prize that compel us to take greater risk? Why do marathon runners post a better time when they run against athletes who are faster than they are? What compels an artist to sacrifice a regular paycheck for the opportunity to stake new ground? Compels us to push past our limitations? To take action in the absence of immediate danger?

Again, we turn to evolution for the answer.

When incentives align with our natural dispositions, the result is always successful. It doesn't matter whether we're talking about flying across an ocean or climbing the corporate ladder, the human organism is naturally driven to acquire influence, resources (money), skills, knowledge, power, and admiration. That's because the ability to overwhelm our enemies, the knowledge to outsmart competitors, the affability to engender empathy, the power to compel others... these advantages enhance our opportunities for success.

And when it comes to swaying our genetic proclivities, there's no question prizes pack a big punch. They motivate us to strive, stretch, and triumph against all odds. To adapt quickly. To leverage our powers of foresight. And to advance society in a way that nothing else has consistently proven to.

The X Prize

In 1996, American engineer and physician Peter Diamandis put the idea of *bioleverage* to the test. The more he studied the role prizes played throughout human history, the more convinced he grew that similar incentives could be used to stimulate competition for the greater good. After all, it was the French Academy's F100,000 (francs) reward in the early 1800s which spurred engineer Nicolas Leblanc to produce soda from seawater—a discovery which gave birth to the chemical engineering industry. And the British Parliament's Longitude Prize was the impetus for John Harrison to perfect the chronometer, a revolutionary timekeeping device to which we owe modern shipping.

In addition to it being a springboard for new industries and disciplines, Diamandis made a second observation about the power of prizes. In each case, *up to one hundred times the actual prize money was invested in the pursuit of victory.* That didn't include the countless numbers of man-hours poured into every failed endeavor in pursuit of a reward. When viewed from this perspective, the return on investment from prizes looked far better than the returns venture capitalists, financial institutions, and sharp investors expected. Put simply, prizes are entirely contingent on success. No success, no prize. No prize, no expense. No expense, no exposure.

There was also a third reason Diamandis became obsessed with contests. As problems grow more complex, unwieldy, and costly to solve, very few corporations, governments, or universities have the financial means to take on long-term, failure-prone discovery—let alone fund multiple teams working in tandem. The cost of primary research has become so expensive it's more than any single institution can bear. But when lucrative prizes are offered, the cost of research and development is spread across many hungry entrepreneurial teams, greatly improving the odds that a successful outcome will be achieved.

Armed with these insights, in 1996—on the forty-third anniversary of Alan Shepard's first spaceflight, and at Diamandis' urging—Anousheh and Amir Ansari announced the first X Prize (named the Ansari X Prize): a $10 million reward for the first team to launch a manned vehicle into suborbital space two times in two weeks.

Twenty-six teams came out of the chute. Some were backed by wealthy corporations and benefactors, while others were self-proclaimed hobbyists. Some were in the middle or tail end of building a viable craft, while others began with a sheet of blank paper. Some understood the magnitude of the challenge, while others were about to find out. Regardless of how big their bank accounts, or how far along their blueprints, one thing was certain: every team was driven by the conviction that their approach was the right one and they would emerge the victor.

Then, in 2004, Burt Rutan laid claim to the $10 million prize. Financed by Microsoft co-founder Paul Allen, the craft Rutan named *SpaceShipOne* completed two flights in suborbital space in one week.

But there was more.

As if breaking previous records for spaceflight was not enough, Rutan also proved spaceflight could be done economically. At that time, NASA's Space Shuttle program cost taxpayers $200 billion, which translated to roughly $1.5 billion per flight. Paul Allen had invested $25 million in *SpaceShipOne*:

a vehicle that was not only lighter, faster, and easier to refurbish, but cost less than 15 percent of a single space shuttle flight.

Foreseeing the future of space travel, billionaire Richard Branson wasted no time purchasing the rights to *SpaceShipOne*. He quickly launched Virgin Galactic and began building a fleet of spacecraft along with the world's first "Spaceport" facility in New Mexico. Virgin Galactic tickets went on sale for $200,000 a passenger for the once-in-a-lifetime opportunity to travel seventy miles above the Earth's surface. And three years after the first X Prize, Virgin Galactic amassed more than $200 million in advance deposits.

Similar to Lindbergh's gift to commercial aviation, Rutan's craft heralded a new era in commercial space transport. By 2012, more than one hundred companies were building commercial crew and cargo transport vehicles as well as space stations, propulsion and launching apparatus, rovers, probes, and space mining equipment. From Denmark to Tokyo, new businesses were springing up—all hoping to grab a piece of the next big thing.

Would Rutan or one of the other teams have eventually succeeded without the X Prize? Probably. Would Michael Phelps have beaten eight world records if there were no Olympic medal? Perhaps. Would engineers labor just as hard were there were no stock options? Maybe. Even if we believe innate ambition is sufficient to motivate first-movers, it would be difficult to deny the positive impact prizes play, especially when it comes to that final push.

Broadening the Bounty

Once the Ansari Prize proved fruitful, Diamandis went on a hunt for patrons to sponsor other prizes. He had a bioleveraged model that worked and there was no limit to the challenges he could apply the model to (*Adaptation Principle Four: success warrants imitation*). Diamandis could foresee a time when a prize for every problem that ailed humanity was offered. Why not?

Then in 2010, Wendy Schmidt, wife of Google CEO Eric Schmidt, stepped forward with a prize that was near and dear to her heart. Outraged to learn the same oil-skimming technology used to clean up the 1989 *Exxon Valdez* oil disaster was being used twenty years later to contain the spread of oil from the Deepwater Horizon disaster, Schmidt started her own investigation. But the deeper she looked, the more disappointed she became. Oil-skimming technology was so inadequate and outdated, experts estimated that, at best, only 3 percent of the five million barrels of surface oil from the Deepwater Horizon leak could be captured. Even working overtime, the skimmers were no match for how much and how fast the oil was spreading. Computer models began showing the extent of the expected damage: currents would soon carry the oil around the tip of Florida, up the eastern seaboard, dealing a potentially irreversible blow to ecosystems along the way.

So Schmidt joined forces with Diamandis to sponsor an X Prize of $1.4 million for the first skimmer that could collect surface oil three times faster than previous methods. And overnight, 350 teams around the world submitted their ideas.

In October 2011, the Elastec/American Marine company of Illinois won Schmidt's X Prize, recovering 4,670 gallons (17,677 liters) per minute. To put this in perspective, the technology used during the Deepwater Horizon crises skimmed surface oil at the rate of approximately 206 gallons per minute. If Elastec's technology had been around at that time, 65 percent of the spill, instead of 3 percent, would have been recovered, making an incalculable difference to the affected communities. In short, the X Prize improved oil-recovery technology by more than 2,000 percent.

Given these facts, it would be difficult to imagine any better way to leverage a $1.4 million investment—particularly when compared to the estimated $40 billion BP—the owners of Deepwater Horizon—spent to clean up the spill.

As word spread about the impact big cash rewards were having, businesses began lining up to inspire similar breakthroughs. It was a cheap way to tap research and development teams around the world at no cost until and unless one succeeded. Qualcomm and Nokia came forward with a prize for the first wireless device for monitoring and diagnosing a person's health that fit in the palm of a human hand. Progressive Automotive offered a purse for a safer, more fuel-efficient vehicle. And before Diamandis knew it, prizes for 100-percent recyclable plastics, clean water, efficient batteries, solar pavement, and electric aircraft were offered.

Stewart and Marilyn Blusson, the sponsors of the Archon Genomics X Prize for the fast sequencing of one hundred genomes explain:

> We believe the X Prize models a new paradigm of philanthropic funding, which not

only provides an incredible amount of leverage but is available to researchers anywhere
in the world operating outside traditional
research and establishments and programs.

The success of the X Prize Foundation is just one example
of leveraging man's natural proclivities to manipulate the outcomes we desire. When social progress slows, when biological
obstacles seem insurmountable, when we struggle to use the
knowledge, technology and resources we have at our disposal
to get ahead of adversity—we must look to what drives the
human organism.

For it is man's innate tendencies that hold the key to acting
on knowledge we possess about the future.

Beyond Carrots and Sticks

While competing for prizes represents one surefire
example of bioleverage, there are many ways to use what
we know about human predispositions to overcome difficult challenges. That said, human nature is complex. We are
compelled by Paleolithic emotions just as often as we allow
the logical part our brain—the frontal cortex—to take the
wheel. So while carrots and sticks work much of the time,
it's not that simple. Successful bioleverage requires us to look
beyond Pavlovian incentives to *the full spectrum of proclivities*
that drive modern man.

Six years ago I came face to face with a problem that showed
all signs of being resistant to either reward or punishment—a
rare case where incarceration and even the threat of death
failed to alter human behavior.

In the summer of 2009, the town of Salinas, a small agricultural community in the central valley of California, was running neck and neck with Oakland, Compton, and Richmond as the most violent city in the state. The Norteños (Northeners) and Sureños (Southeners) gangs had declared Salinas their home base, and the result was open warfare. Salinas' homicide rate was running five times the national average. Even more daunting was the fact that out of a total population of 145,000, nearly 4,000 had been identified by law enforcement as violent gang members. At that time, Salinas employed roughly one officer for every eleven thousand residents. So, assigning more than half an officer to gang activity left the other residents of the city vulnerable.

The situation was grave. You didn't need predictive analytics to see where the city was headed.

One afternoon the mayor of Salinas invited me to meet him, the county sheriff, and a recently retired judge to discuss the rise in crime. The irony of the meeting place the mayor chose was difficult to ignore. In 1873, a lynch mob overpowered the local jail and strung Matt Tarpy from a tree for murdering Sarah Nicholson over a land dispute. Since that time, the site of the hanging had become Tarpy's Roadhouse, an elegant restaurant with a lovely garden and dark past. Though vigilante justice might have been commonplace in the 1800s, I hoped it wasn't on the menu that afternoon...

After the mayor made a few short introductions, the three men began explaining where the violence was occurring, how the police were responding, the initiatives that had been tried (a gang taskforce, neighborhood watch groups, etc.), and what they had learned about gang culture. In particular, I was interested in the sheriff's perspective, because his organization and the police were often first on the scene.

> We know where the problem areas are. But basically what happens is the police get a call that someone's been shot and then they send officers over to collect evidence, and that's it. We don't have the time or manpower to investigate every murder, and since the violence is in one area, and mainly between gang members, we try and stay clear of that area. We're letting them kill each other. Look. It's dangerous for an officer to go in there. They need backup. Otherwise, they pull someone over, and in minutes they're surrounded by other gang members, who'll be standing around the officer, pressing in, shouting, and many times there's a problem before backup can get there. The officer's outnumbered. They can't arrest everyone...its dangerous.

The sheriff made a point to tell me that not only were the police outnumbered, the gangs were also better financed and equipped. They had access to military-grade bulletproof vests and powerful, fast-firing weapons. In contrast, with state and local budgets slashed to the bone, a police officer could wait months to get their hands on a stun gun.

Then the retired judge jumped in.

From his perspective, the problem was jobs. The recession had eliminated jobs for individuals who didn't speak English and had marginal skills. Organized gangs had stepped in to fill that need. They offered immediate, paid employment. And with that employment came respect, friendship, loyalty, a hierarchical structure with a clear pathway for promotion, bonuses, and other benefits. Gangs had also turned state and local prisons into executive training camps for members interested in fast-tracking their careers. So sentencing and incarceration no longer worked the way it was supposed to. Many gang members wanted in.

I listened carefully as the mayor, judge, and sheriff all painted a similar picture: the gang leaders were functioning like successful entrepreneurs—expanding territory, recruits, infrastructure, and market share faster than the police and judicial system could slow their roll.

When lunch was over, I was sure I was in over my head. What did a sociobiologist know about gang warfare and law enforcement? But in the days that followed, the weight of young men gunning each other down in the streets bared down on me. So did reports of children hit by stray bullets, injured officers, and news that eleven-year-olds were being lured into gang life. There was a way to use bioleverage to reverse the threat, but I did not know what it was...

Weeks later, I was standing in my kitchen drinking my morning coffee, watching one hummingbird battle the other around the feeder outside, when it hit me: to win the battle against gang violence *authorities must leverage human nature, not try to change it.* The key to restoring law and order in Salinas was to first acknowledge that, at the present time, *the human organism is designed to protect three things*: itself, others it has a connection with, and territory. That's it. We'll risk life and limb to defend the same things a hummingbird will.

Early in our history, military leaders discovered this phenomenon and used it to their advantage. They learned that clear territorial boundaries and troop bonding were essential to victory. Sun Tzu, Caesar, Napoleon, Churchill, and MacArthur all recognized the value of solidarity between a troop and its leader—between soldiers and those they were ordered to defend—as well as securing prominent landmarks. In fact, the link between troop, territory, and victory is so profound, professional sports later adopted these protocols. This is how "zone" and "man-to-man" coverage were born. Similar to soldiers, players perform better when they have specific areas and individuals to protect and defend against.

But when it came to Salinas law enforcement, every officer was tasked with protecting *all of the citizens, all of the time, in every area of the city.* There was no mechanism for officers to bond with the people they were asked to risk their lives for. Or specific territories or boundaries to defend. According to the sheriff, when a report came into the station, officers closest to the scene were ordered to respond.

No wonder the city was losing ground. No "zone" or "man-to-man."

There was a second reason law enforcement was struggling to gain control. Gang kingpins has assumed the "alpha" role in the community and in that process demoted the police to "beta" status. The new alphas offered steady employment; they restored member's self-worth, purpose, and identity; they provided protection; they administered swift justice when needed; they even looked after a convicted offender's family when they were injured or serving time in prison. What did a police officer who showed up to investigate a homicide after the fact, was not from their neighborhood, and didn't speak Spanish have to offer?

The more I thought about it, the more Salinas' situation reminded me of Times Square before Giuliani took over.

In the early '80s, Times Square had descended into an inner-city slum known for crime, drugs, prostitution, and homelessness—a place tourists were advised to avoid. Yet Times Square was also the most famous place in the world to celebrate New Year's Eve, home to New York's acclaimed theater district, and where the city's top luxury hotels were located.

In its heyday, Times Square had been the heart of New York City.

With this in mind, Rudy Giuliani's first order of business as mayor was to resuscitate Times Square. He was determined to bring it back to its former grandeur—and do it quickly.

On day one Giuliani instructed the precinct commander to install an officer on every block in Times Square. Two if needed. With financial help from the federal government, New York City undertook one of the largest law enforcement recruitment campaigns in US history, hiring more than 3,660 new recruits. Giuliani also instituted the now famous CompStat program, a system that allowed crime activity to be collected and analyzed twenty-four hours a day.

Every morning, precinct captains began their day by reviewing the previous day's activities—examining how every street, block, and officer had performed. Since every inch of Time Square was assigned to a specific officer, officers were now accountable for everything that happened on their block during their shift. If there was an incident in the officer's assigned territory, the precinct captain knew about it by the following morning.

Under Giuliani's initiative, police officers were ordered to be visible—not only to potential perpetrators, but to merchants and customers who were conducting business in the area. Police officers were trained to check in with shopkeepers at the beginning of each shift and to interact with the public.

It didn't take long before business owners knew the officers' names and schedules and vice versa. Shopkeepers began offering the officers coffee and free food, sharing pictures of their children, and telling them about their family vacations. They were grateful for police protection and began bonding with their protectors. Likewise, officers were soon responsible for more than an assigned block, but for people they saw every day: the mother, father, student, or sister they came to know.

But Giuliani didn't stop there.

Officers were ordered to enforce a zero-tolerance policy in Times Square. If a tourist so much as dropped a candy wrapper on the ground, an officer would politely ask them to pick it up and point to a nearby trash can. There was to be no litter, no jaywalking, no loud boom boxes, no panhandling permitted. This was *the officer's block*, and it was up to them to exert control.

And it worked.

The more law enforcement cracked down, the faster their alpha status returned. Though many citizen groups cried foul because the police were fining people for minor infractions, such as loud music, the mayor knew the importance of establishing dominance. He wouldn't give an inch.

Giuliani and the NYC police department cleaned up Times Square in a few short years. While crime across the United States decreased by 30 percent during this period, it fell 56 percent in New York City under Giuliani's take-no-prisoners plan. And Times Square—an area that saw over four thousand crimes in 1993—became one of the safest, most vital districts in New York City. Tourism returned. So did businesses like Disney, Nike, and Bank of America. And today, the streets of Times Square are so crowded, every cab driver knows better than to go anywhere near Broadway and Seventh after 9:00 a.m.

Based on Giuliani's success with Times Square, my recommendations to the mayor of Salinas were similar. Put a beat cop on every block of known "hot zones" in Salinas—two and three where violence was escalating. If federal and state funds weren't available to hire more officers, raise funds through citizen action groups and business associations. The mayor was going to need more bodies. A lot more.

Officers would be instructed to go door-to-door when they came on duty, announcing to every household and business in their territory they were there to assume the next shift. It was important for the police to be friendly and accessible and build a connection with the people they were assigned to protect. In turn, the community would begin to reciprocate with acts of kindness—just as appreciative business owners in Times Square began offering officers coffee, shelter when it rained, and friendship.

The Salinas police force would have to enforce a strict, zero-tolerance policy in order to reclaim their alpha status. If so much as too many vehicles were parked in a driveway, they must issue a citation and seek the maximum punishment or fine allowed by law. This is where the retired judge could help—by soliciting the full cooperation of the judicial system.

To further relegate gang leaders to beta status, I urged the mayor to deploy one of the most powerful and under-utilized enforcement tools in nature: humiliation. Among primates, alpha males and females often keep challengers in check by using humiliation, ridicule, and shame, rather than resorting to physical intimidation. So, though unorthodox and tainted by an abusive history, I recommended instituting chain gangs.

But in this case, it was important not to assign chain gangs to hard physical labor, because physical labor was a sign of strength. These chain gangs must be assigned to menial public tasks—tasks unfit for alpha males and females, such as sweeping streets with a push broom, cleaning public toilets, collecting trash, or handwashing police vehicles. Furthermore, the chain gangs must be put on public display in areas where school buses passed in the morning and afternoon, in the neighborhoods where the offenders lived or conducted business, and in heavily populated areas such as shopping malls, sporting events, etc. Any attempt to glamorize gang and prison life needed to be quashed to inhibit future recruiting.

But there was more.

To regain alpha status, it was important for law enforcement to do more than protect. They must also provide. Since employment was one of the primary motivations for joining a gang, unless there was a pathway for gang members to assimilate into mainstream society, it was likely these measures would fall short. People need work. They need to eat and provide for their families. If their only viable employment is a violent gang, then gang it is.

Subsequently, the mayor consulted the head of the city's gang taskforce, who agreed the root of the problem was unemployment. But he had an idea... one the mayor could immediately deploy.

Since Salinas had a policy of awarding city contracts to local companies, why not embed a requirement to hire, train, and utilize parolees who were serious about learning a trade? In turn, to qualify for a trade training, a gang member would be required to enroll in a GED program, agree to weekly drug testing, and see a counselor who would assist them in breaking gang ties and assimilating into mainstream society. According to the head of the gang taskforce, gang members would choose a legitimate job over gang life any day of the week. So once the program proved to be successful and word spread, gang members would defect one after another on their own.

Armed with this knowledge and the tools to bioleverage, the mayor, sheriff, and judge went to work. They obtained federal funding to initiate a gang job retraining and reintegration program. They began identifying at-risk youngsters and working with schools and organizations like the Boys and Girls Clubs of America to offer these children the guidance and support needed to assimilate and succeed. Salinas hired more police officers and sheriff deputies and assigned them to specific physical territories. And over time, the number of homicides began to level out. Then slowly Salinas turned a corner. Murder rates fell by 40 percent the following year. And they remained at that rate the year after that, despite an influx of gang members. And just when things were headed in the right direction and law enforcement was getting a grip, the residents of the city decided the crackdown was more than they bargained for.

And just like Giuliani, the mayor and sheriff were voted out of office.

Both Selfish and Selfless

Most of us recognize the fact that the drive to survive is fundamental to living organisms. Every living creature is programmed to perpetuate their genetic legacy, and this has a strong bearing on behavior. But recently this fact has cast doubt on whether selfless acts are selfless at all. Are all life-forms driven by a sort of me-first genetic narcissism? And does that mean bioleverage only works when it appeals to selfish motives?

In 1964, W. D. Hamilton put forth such a theory—one which ascribed mercenary motives to altruism. Widely embraced by the scientific community, Hamilton's theory of kin selection asserted that when we are not acting to insure the continuation of our own DNA, we are acting to protect subsets of our DNA found in distant—sometimes remote—kin. In other words, what appears like altruism is really just a way of looking out for ourselves by looking out for diluted versions of ourselves.

Not exactly a flattering picture of humanity...

But biologist and naturalist Edward O. Wilson didn't quite see it the same way. In an article about Wilson in the *Boston Globe*, Leon Neyfakh writes:

> Researchers were finding species of insects that shared a lot of genetic material with each other, but didn't behave altruistically, and other species that shared little and did. "Nothing we were finding connected

[altruism] with kin selection," Wilson said.
"I knew something was going wrong—there
was a smell to it."

So Wilson, together with Harvard University's Biology and Mathematics expert Martin Nowak, proposed an alternative theory: altruism, teamwork, sacrifice, and tribalism were strategies designed to benefit the survival of the group—not kin—and sometimes to the grave disadvantage of the individual.

This was a radical notion. It meant that we are not only designed by nature to act in our own self-interest, but also on behalf of "group selection."

In other words, we are both selfish and selfless. And it turns out, so are other socialized creatures.

In his Pulitzer-prize winning book *On Human Nature*, Wilson explains how animals regularly expose themselves to risk for the greater good:

> Certain small birds, robins, thrushes, and titmice, for example, warn others of the approach of a hawk. They crouch low and emit a distinctive thin, reedy whistle. Although the warning call has acoustic properties that make its source difficult to locate in space, to whistle at all seems at the very least unselfish; the caller would be wiser not to betray its presence but rather to remain silent.

It turns out, acts of altruism and cooperation are quite common in nature. For instance, social insects—such as ants and bees—regularly commit altruistic suicide against intruders. Among chimpanzees, Jane Goodall observes that orphaned infants are routinely adopted by the brothers and sisters of the deceased. And even among vampire bats—one of the last species in which we might expect to observe benevolence—a form of "reciprocal altruism" regularly occurs. Author Robert Wright observes:

> Since blood is highly perishable, and bats don't have refrigerators, scarcity faces individual bats pretty often...bats that return to the roost empty-handed are often favored with regurgitated blood from other bats—and they tend to return the favor in the future.

What do these and other examples of goodwill tell us? It is just as important to leverage our genetic proclivities to cooperate, collaborate, and sacrifice as it is to appeal to self-benefitting motives. After all, humans owe much of our evolutionary success to troop life. Troop collaboration allowed us to fend off predators larger and fiercer than any single individual could, and to enjoy the many advantages of cooperative hunts. Teamwork provided additional protection throughout a long gestation period, and also for our young—allowing our numbers to grow. It was also responsible for the development of sophisticated oral and written communication, facilitating the passage of knowledge from one generation to the next. There's a reason ants, bees, and humans take the prize for group selection—and that reason is cooperation.

The Collaboration Gene

One day I decided to see if I could bioleverage our genetic predisposition to collaborate and work as a team. I worked as a part-time instructor at a local university, and every week I watched as students filed in, took a seat, and immediately jumped on their phones and tablets without so much as saying a word to the student sitting next to them.

It was the perfect setting to see what would happen if they were assigned a task which could not be completed on their own.

My experiment was simple. Once the students were situated, I passed out a multipage survey. Then when every student had a copy, I explained that there were clear and specific instructions on each page, and they must follow the instructions *exactly* as they were written. And most importantly, they must answer *all* the questions on every page to receive class credit. Then I apologized for having to leave for a few minutes to take brief phone call. As I was headed for the door, I mentioned the department copier had been acting up so some of the copies might be a little cockeyed or a lighter shade of ink, but do the best they could.

Then I left to take my call.

What the students didn't know was that most surveys contained at least one page which was cut off, smeared, or completely illegible. Only a small handful of students received documents where all pages were legible.

For the first few minutes the students worked quietly on their own—filling out what they could. Then, after a short while, they began asking students to their immediate left and right about what was on their damaged page. It didn't take long before the entire classroom was noisy with students helping each other.

Then finally, an alpha male stood up and said, "Whoever can read the questions on page five needs to read them out loud for everyone." And a young woman in the back of the room obediently got up and read the page aloud for the entire class.

This was followed by other students standing and reading questions on other pages.

By the time I returned to the classroom, all of the surveys were neatly stacked at the front of the room and students were busy conversing. The room was animated, lively, noisy. It was not the same classroom I had observed for weeks.

So I apologized again for having to take a call and asked the students, "Anything happen while I was out?"

No one answered.

So I asked again. "What happened? Sounds like I missed something."

They looked confused. The same alpha who had ordered the girl to read the page out loud shook his head. "Nothing. You asked us to fill out the survey so we did."

"No, you didn't." I smiled.

"Yeah, we did. They're right there." He pointed.

"No. I think you did something else... You *collaborated*."

Up until the survey, it was clear every student felt confident they could pass the class without the help of their peers. But given a task we cannot complete by ourselves, and a limited time frame, we are programmed to react the way shipwrecked passengers or strangers lost in the wilderness do—we quickly organize for our collective survival. Funny how pride, racism, sexism, and ageism become luxuries we cannot afford when our survival is at stake.

After the survey, I noticed some students began working together on other difficult assignments. So, in order to engender more collaboration, I decided to allow students to ask each other for assistance during exams. They were not permitted to use their computers, cell phones, books, or any other resource, but they could confer with a fellow classmate. In quick order, groups of three, four, and five began sitting together during tests. Students who thought this was reason not to study and contributed nothing soon found themselves ostracized for not pulling their weight, whereas better students in the class seemed to have no shortage of friends. At one point I could almost guess a student's grade by whether they sat alone or in groups, and how big the group was.

Which brings me to the point of my ad hoc experiment. The rewards, discouragements, and conditions we engineer don't have to be as big as the X Prize. They don't have to be as extreme as the measures Giuliani or the Salinas police force undertook. All bioleverage requires is an understanding of human nature.

Today, when I am invited to work with corporations whose ability to innovate or forward momentum has stalled—when I witness siloing, territoriality, fear, and competition rather than collaboration—I revert to my classroom experiment. I ask executives to perform impossible tasks within an unreasonable timeframe and draw upon their natural instinct to band together during times of duress. And from there we build. We use foresight and analytics to chart a clear course with known outcomes. Identify *biobstacles* that may prevent us from implementing. Then use the principles of adaptation and bioleverage to move swiftly toward our objective— tipping our hats to previous limitations as we pass them by.

"A serious prophet, upon predicting a flood,
should be the first man to climb a tree.
This would demonstrate that he was indeed a seer."

- Stephen Crane,
The Red Badge of Courage

CHAPTER EIGHT

A Prescient Mind

Not long ago I read a story about a woman whose home was situated next to a busy highway. During rutting season, it was common for cars to strike deer which were crossing to the other side. The injured deer would stumble onto her property and die. So she complained to the town council that every year she had to pay someone to remove the dead animals and clean up afterward and the town was to blame.

When the mayor asked why the town was responsible for the dead deer, she answered, "I've been here many times and you won't listen! If you would just move that Deer Crossing sign a little farther up the road, they wouldn't end up in my yard!"

My friend Luis Alvarez is fond of saying, "A funny thing happened on the way to the future." That funny thing is how our brains adapted. Or in some cases, didn't.

To this point we've looked at how strategic and tactical adaptation (bioleverage, predaptation, predictive analytics, and other methods) can be used to capitalize on foresight. But what about human physiology? Are our brains becoming better predictors? Better adapters?

The short answer is yes.

Though social adaptation is faster, more efficient, and generally more successful than physical evolution, our bodies continue to advance in ways that are mysterious and impressive. And nowhere do we see greater evidence of this than in neuroscientific studies that show our brains are better at anticipating, and more malleable, than we knew.

Predictive Neurons

Neuroscientists and behavioral psychologists have long understood the role naturally occurring chemicals in the body—such as serotonin, oxytocin, norepinephrine, phenylethylamine, dopamine, and hormones like estrogen and testosterone—play in how we learn and make decisions. But what we didn't know until recently is there is a special class of neurons in the brain that use this chemical reward system to make predictions. Scientists call them "prediction neurons" and their purpose is to make accurate guesses about future events.

Dr. Wolfram Schultz, neuroscientist at the University of Cambridge, is largely responsible for the discovery of prediction neurons. He devised a simple scheme to test whether our chemical feedback loop rewards foresight: Test monkeys were exposed to a loud sound, after which an apparatus squirted sweet apple juice into each monkey's mouth. The sound-squirting sequence was repeated many times so the monkey had an opportunity to associate the sound with the juice. At first blush, Shultz's experiment appeared to be nothing more than an exercise in positive reinforcement—similar to Pavlov's early experiments with dogs. But what made Schultz's research noteworthy was evidence that the monkey received a boost of dopamine when its expectations were correct.

Here's what happened...

At the beginning of the experiment, the monkeys' brains released dopamine at the same time they received the squirt of juice. But soon, the timing of the dopamine shifted. The brain began releasing dopamine *prior* to the juice—rewarding the monkey for correctly assessing what was going to occur *next*. Later, when the experiment was modified so some sounds were no longer followed by juice—and the monkey's expectations were frequently incorrect—the amount of dopamine sharply declined. The monkeys were punished for wrong guesses about the future.

Prediction neurons work the same way in the human brain. The area of the brain responsible for imagining future events is called the anterior cingulate cortex (ACC). When we fail to anticipate correctly, neurons in the ACC dispense chemicals that make us feel surprised, irritated, anxious, frustrated, defeated, even angry. These emotions are common when we fail to get that promotion or raise we expected, when we encounter traffic we haven't planned for, when an unexpected bill arrives in the mail, when a spouse surprises us with a divorce. Our bodies admonish us for failing to see what was coming.

And by the way, this also includes "good surprises." Which explains why many people report feeling irrationally fearful, anxious, and overwhelmed when they open the front door to friends and family yelling "Surprise!" or when a loved one suddenly pops the question, or when strangers arrive with a camera crew to deliver an oversized sweepstakes check. Prediction neurons don't discriminate between good and bad surprises. Their primary role is to foresee the future to give us time to prepare. And in nature, as in government and business, that advantage may make all the difference to our survival.

Prophetic Propensities

Owing to an adroit chemical reward system, humans are designed to make predictions all day, every day. We have a pretty good idea who is at the front door before we get there, what we are getting for our birthday, and whether we will be asked for a second date or interview. We know when we're going to be hungry, when we'll need to stop for gas, when we'll run out of money, and, if you're like me and travel a lot, when your flight is going to be delayed because the plane hasn't arrived at the departure gate yet.

But what's truly astounding is how often our predictions are spot on. Think about this. We make thousands of predictions every day with nothing more than our brain's ability to deduce and create imaginary scenarios. No Big Data systems, no analytics, and in most cases, very little empirical evidence to go on. And that's impressive.

So impressive that Jeffrey Zacks, researcher at Washington University, decided to put the brain's prescience to the test. He began showing videos of simple events to a group of students. Then he stopped the videos midway and asked his students to write down what happened *next*.

Over 90 percent of the students gave the correct answer. Ninety percent.

Meanwhile, at the University of California at Los Angeles, professors of psychology Matthew Lieberman and Emily Falk, wondered whether brain activity was a good predictor of future behavior. To find out, they designed a simple experiment. The duo recruited subjects who had never used sunscreen. Then each subject's brain activity was monitored as they were exposed to a series of public service messages about the importance of protecting skin from the harmful effects of ultraviolet rays. They soon discovered that individuals who showed increased activity in the medial prefrontal cortex while watching the public service messages showed a much greater use of sunscreen afterward than those who showed little or no brain activity in this area. In other words, by observing brainwaves alone, they could accurately predict what a person was likely to do.

Kent Kiehl, a neuroscientist, at the Mind Research Network, took this idea further. He wanted to know whether brainwaves could be used to predict whether prisoners would reoffend in the future. In Kiehl's experiment, prisoners were asked to view a computer screen where either the letter K or X would appear. They were instructed to quickly press a button *only when the letter X appeared*. But do nothing when they saw the letter K. The "test was rigged so X popped up 84 percent of the time,"—which predisposed the prisoners to expect an X. This made it difficult to refrain from pressing the button on the rare occasions when a K appeared.

The result? The area of the brain responsible for impulse control turned out to be the most accurate predictor of criminal activity we have discovered to date:

> Inmates with relatively low anterior cingulate activity were roughly twice as likely as inmates with high anterior cingulate activity to be arrested for a felony offense within four years of their release.

According to Essi Viding, a professor of developmental psychopathology:

> Interestingly this brain activity measure appears to be a more robust predictor, in particular of non-violent offending, than psychopathy or drug use scores, which we know to be associated with a risk of reoffending.

And clinical psychologist Dustin Pardini added, "It's a great study because it brings neuro-imaging into the realm of *prediction*."

Prediction, indeed. Until these studies, we had no idea whether brain activity could be used to pinpoint future activity. But it was only after scientists at Northwestern University suggested that brain activity could predict acts of terrorism that the public took notice. If prisoners foreshadowed their behavior, why not organized terrorists?

P300

To test this theory, individual students were asked to plan a terrorist attack. They were instructed to write down as many details as possible about their attack, and to put the detailed plan in a sealed envelope. The students were prohibited from discussing or showing their plans to anyone until the experiment was over. So, the architect of each plan was the only one who knew when, how, and where the attack would occur—no one else.

Then each student was asked to look at the written names of major cities while their brain activity was audited. Next they were shown the names of weapons. Followed by the name of every month of the year. And lastly, dates of the month and hours of the day. The scientists were looking for a particular brain wave called P300—a marker "associated with feelings of guilt and secrecy." And just as they anticipated—with no clues to go off except for the strength of P300 readings—researchers were able to determine the city, weapon, and exact day and time of planned attacks greater than 83 percent of the time.

The ramifications of this result are difficult to ignore. When compared to the low success rate associated with torturing suspects to extract information about terrorist events or acting on unverifiable data provided by questionable ground assets, 83 percent is an extraordinary achievement. The fact is, brainwave technology is not only a faster, more accurate method of identifying and preventing future threats, there is no danger of overstepping the law or acting on dubious data.

In truth, neuroscientists have made remarkable inroads when it comes to understanding how our brains make predictions. And with a flood of new technologies and predictive algorithms on their way, our prognostications will only improve with time. Until one day, the future will no longer be the future. It will simply be one of several known options we choose from.

Fast Evolution

No discussion about the hidden capabilities of the human brain would be complete without talking about the controversial London black-cab driver study—a study that calls into question everything we know about how evolution works.

In nature, rapid mutation is generally associated with severe changes such as an ice age, extreme drought, a drop in oxygen, rampant disease, and so on. And more often than not it occurs over several generations—spanning a longer period of time than is optimal. For this reason, extinctions are common. There is no such thing as physical adaptation at a moment's notice.

Or is there?

Doctors Eleanor Maguire and Katherine Woollett from University College London set out to find out. And they chose as their subjects the drivers of London's black taxi cabs.

If you do not live or work in London, then you have no reason to know that, on average, black-cab trainees study for four years and make twelve or more attempts to pass the infamous Knowledge of London Examination System required to become certified. It is one of the most difficult tests in the world. The exam requires trainees to be familiar with more than twenty-five thousand streets, as well as thousands of obscure British landmarks dispersed throughout a tangled, dense nest of alleys, one-way streets, dead ends, and government thoroughfares. Drivers must demonstrate a comprehensive understanding of traffic patterns throughout the day, where tolls must be paid, where taxis are permitted to wait or park throughout the city, including at all landmarks, and on and on and on. So not surprisingly, less than half who attempt the Knowledge of London Examination System ever pass. The other half walk away.

Due to the degree of difficulty involved, in 2000, Maguire and Woollett began examining the brains of seventy-nine driver trainees. Using magnetic resonance image (MRI) technology, they scanned the brains of the trainees and those of thirty-one control subjects who had no interest in becoming a driver. The scans revealed no differences between the brains of trainees and the general public.

Then later, thirty-nine of the original seventy trainees passed the dreaded Knowledge exam. At this point the scientists rescanned the trainees' and non-trainees' brains to determine whether the arduous requirements of the exam had any impact on the brain.

And what they discovered astonished and confounded biologists everywhere.

The posterior of the hippocampus—the part of the brain associated with navigation and long-term memory—had grown measurably larger in the drivers who passed the Knowledge exam.

And what appeared in later scans was even more startling: the longer a cabbie stayed on the job, the larger this part of their brain grew. Drivers who remained on the job for forty years showed the greatest enlargement of this area of the brain.

This was the first time scientists observed a direct task (navigation and recall) changing the development of a brain in a tangible, measurable way. Or as Dr. Maguire put it, "we can now see there *can* be structural changes made in healthy human brains."

Which begs the question: if the Knowledge exam can have an impact on brain development and evolution, what else can we affect? The London black-cab drivers study challenges our previously held views that physical evolution is random and requires a longer time span.

We now know that Mother Nature's hand can be forced. And with that knowledge, the journey to engineer better brains has begun. In April 2013, the president of the United States announced a government-sponsored program to "map the human brain," the Brain Research through Advanced Neurotechnologies program, better known as BRAIN, which experts claim will lead to a greater sea change than the mapping of the human genome. And other industrialized nations, including China, Japan, and the United Kingdom have announced similar initiatives.

If there was any doubt as to whether our capacity for foresight would be limited by human physiology, London's black-cab drivers and the new quest to uncover the hidden potential of the human brain should put that doubt to rest. We are only beginning to scratch the surface.

"It is not in the stars to hold our destiny
but in ourselves."

- William Shakespeare

CHAPTER NINE

ForeWorld

*On January 6, 2015, humankind passed an important milestone.
Astronomers catalogued the one-thousandth planet in space. That
same day, scientists at NASA announced the Kepler spacecraft had
"compiled a list of 3,500 more candidates."*

Thirty-five hundred more planets?

The hunt for an Earth-like twin is on.

*The latest estimate is there may be as many as 40 billion "habit-
able, Earth-sized planets in the galaxy." As this number continues
to grow, the odds of discovering extraterrestrial life have graduated
from a plausibility to a probability. It's now a matter of time.*

*The announcement from NASA got me thinking about how our
economic and political institutions, religious beliefs, customs, and
animosities might be affected by contact with extraterrestrial life.
I pushed and shoved and banged against the outer limits of my
ability to project ahead: could we use foresight to prepare for the
inevitability? Predapt?*

Well, for openers, there's a big difference between discovering single-celled organisms versus creatures that are one, two, three thousand years ahead of us. Finding bacteria, protozoa, or algae isn't likely to have much impact on our daily lives. On the other hand, highly advanced life-forms would find us no more amusing than protozoa. They'd have about as much interest in trying to communicate with humans as we do with germs and insects. This means, either way, we have nothing to fear.

I found this thought comforting. It meant humankind was still in the driver's seat. Still in charge of our destiny...

Meanwhile, as I was contemplating the future of our tiny outpost in the universe, a group of particle physicists on the other side of the world were using foresight to explain the future of reality. You, me, the stars and planets, dark matter, everything.

The Higgs field put a frightening new spin on destiny. According to scientists at the European Organization for Nuclear Research, the only thing preventing atoms—which make up everything—from zooming through space at the speed of light is a particle that interacts with every atom to give it mass and slow it down. Atoms are running through a thick, soupy swamp which slows them down just enough to create the physical universe. Our bodies, our cars, our dogs, the ground we walk on, and the stars overhead—all courtesy of the Higgs field.

According to science writer Dennis Overbye, this puts us in a precarious situation:

> *The idea is that the Higgs field could someday twitch and drop to a lower state, like water freezing into ice, thereby obliterating the working of reality as we know it. Naturally we would have*

> no warning. Just blink and it's over...Talk about
> a particle with Godlike properties.

"A particle with Godlike properties." A particle on which I depend to wake myself up, get my cup of coffee, pack the kids off to school, and transport myself to work. A particle that can only be mathematically described. One that will decide my future, along with all that surrounds me.

It turns out, we may not be in the driver's seat after all...

But now that I have this information, what do I do with it? Do I cancel my dentist appointment? Forget about my job, bills, laundry, and sit on the beach, eat a sandwich, and watch the sun set? After all, this could be the last sunset. The last flash of orange and pink for all of eternity. The last time atoms stand still long enough to stir the souls of poets and lovers. Dare I miss the Earth's last goodbye?

The famous French existentialist Albert Camus was more prescient than he realized when he remarked, "Fate is not in man, it is all around us." Like fireflies trapped in a jar, we are all light, waiting for the field to set us free.

Last week a colleague told me the reason the Earth was round was because we're not meant to see too far ahead.

"Too late," I replied.

The future is calling. It's calling on our smart phone, computer, and the next predictive algorithm. It's calling at work, at school, and at the corner coffee shop where a high-speed Wi-Fi connection is waiting.

It's calling.

And we have answered the call.

But there is static on the line, and we can't quite make out all the details. Even when we can, it's unclear what to do about them.

That's the problem with foresight and preemption. Since it's impossible to prove an event which has not yet occurred will occur, any attempt to preempt quickly descends into a debate over whose imaginary scenario is correct. Was the preemptive march into Iraq necessary? How about a police officer who fires on a potentially dangerous suspect? Banning immigrants from countries known to harbor terrorists? Stricter gun control and carbon emission laws? Barring a parallel universe where we do nothing, there is no way to compare—no way to know for certain if preemption was necessary or if it worked. When it does work—and a problem is prevented from escalating or occurring—it only adds credence to the argument there was no problem to begin with. And when preemption fails, well, there's plenty of finger-pointing to go around.

Today, preemption is a lose-lose proposition. No one ever received an award for preventing a problem. There can be no gratitude, no accolades, for something that never happened. To be a hero we must first wait for the building to catch fire. For 9/11. The Avian flu. Fukushima. Aleppo. Global burning.

But that will change.

It will change because the future will no longer bow to speculation. It will answer to data, predictive analytics, and lightning-fast computing and be forged in the fire of predaptation. It will become the providence of those who dare to act on foresight.

The Verge

We stand on the cusp of a decisive transition. One which will forever alter the way governments, businesses, and individuals progress. Foreknowledge and fore-action combine to set us on a new course. Armed with the technology to predetermine outcomes, the future can be shaped to ensure our eternal prosperity. To achieve this promise, we need only grab the reigns of self-determination and act on what we now know—with increasing certainty—lies ahead. We need only embrace a capability so powerful it has been reserved for humankind alone. We need only assume our rightful place as aspiring Masters of the Universe.

"Mankind is a weaver who, from the wrong side,
works on the carpet of time. The day will come when he will see
the right side and understand the grandeur of
the pattern he, with his own hands, has woven through the
centuries, without seeing anything but a tangle of string."

- Alphonse de Lamartine

ACKNOWLEDGMENTS

I have many people to thank for this book and my work. First and foremost, Edward Wilson, whose poetic writing and scientific prowess has inspired me for four decades. Wilson never wasted a minute worrying about his reputation. I wanted to be strong like that. I still do.

I thank Gary Robinson, Kent Hansen, Dennis Donohue, and others who I interviewed, as well as those whose research I drew upon. And Arthur Klebanoff and Hannah, Brian, and Michelle at RosettaBooks for bringing *On the Verge* to life. I also owe a debt of gratitude to the teams at KSCO and APB for extending my reach to radio and public speaking.

Mackenzie Lovelace kept my office running for long periods when I was writing. It was no easy feat, and I remain forever grateful for her support. Marianne Pittard looked high and low to find a cabin off the Oregon coast where I could hear myself think. John Laughton, Jane Marcus, and Andrea Massion always had an encouraging word when my writing stalled. And my son, Mathew, never missed an occasion to leave flowers on mom's desk.

Lastly, I want to thank John Willison for sending me into the world to find my voice. I showed you the Pleiades, and you showed me the way.

NOTES

PREFACE

p. 1 **More than a hundred fifty years ago, Charles Darwin revealed that it is our ability to adapt that is the determining factor**

Darwin, Charles. *On the Origin of Species*. New York: Sterling Publishing, 2008.

CHAPTER ONE
Foresight

p. 8 **–and the failure of over 99 percent of the species that once inhabited Earth**

Stearns, Beverly Peterson, and Stephen C. Stearns. *Watching, from the Edge of Extinction*. New Haven: Yale University Press, 1999.

Novacek, Michael J. "Prehistory's Brilliant Future." *New York Times*, November 8, 2014.

Newman, Mark. "A Model of Mass Extinction." *Journal of Theoretical Biology* 189, no. 3 (1997): 235–252.

Evolution, Television Series. Public Broadcasting System, 2001.

Kolbert, Elizabeth. *The Sixth Extinction: An Unnatural History*. New York: Picador, 2015.

"The Five Worst Mass Extinctions." *Endangered Species International*. Accessed May 8, 2017. http://www.endangeredspeciesinternational.org/overview.html.

Ward, Peter. "Is Earth Undergoing a 6th Mass Extinction?—'99.9% of All Past Species Extinct.'" *The Daily Galaxy* (blog). March 23, 2013. http://www.dailygalaxy.com/my_weblog/2013/03/of-all-species-that-have-existed-on-earth-999-percent-are-now-extinct-many-of-them-perished-in-five-cataclysmic-events-t.html.

p. 9   They created three goddess-like "Fates" called the Moirai to explain the trials and triumphs mortals encounter

Herodotus. *The Histories.* Floyd, VA: Wilder Publications, 2015.

Rose, H.J. *A Handbook of Greek Mythology.* New York: Plume, 1959.

Greene, William Chase. *Moira: Fate, Good, and Evil in Greek Thought.* Cambridge: Harvard University Press, 1944.

Atsma, Aaron J. "Moirai." *Theoi Project.* 2000. http://www.theoi.com/Daimon/Moirai.html.

p. 10   According to the principles of Karma, every living organism is held accountable for their actions

Lopez, Donald S. Jr. *Religions of Asia in Practice: An Anthology.* Princeton: Princeton University Press, 2002.

Thakkar, Chirayu. "Karma." *Ancient History Encyclopedia.* December 4, 2015. http://www.ancient.eu/Karma/.

Reichenbach, Bruce R. "The Law of Karma and the Principle of Causation." *Philosophy East and West* 38, no. 4 (1988): 399–410.

Winichakul, Thongchai. "Buddhism and Society in Southeast Asian History," (University of Wisconsin-Madison, Madison, 2008).

Chodron, Thubten. *Good Karma: How to Create the Causes of Happiness and Avoid the Causes of Suffering.* Boulder: Shambhala Publications, 2016.

p. 10   In areas of Africa and South America sacrifice plays a large role in determining the future

Mbiti, John S. *African Religions and Philosophy.* Oxford: Heinemann, 1990.

Beckwith, Carol, and Angela Fisher. *African Ceremonies: The Concise Edition.* New York: Harry N. Abrams, 2002.

Agorsah, Kofi E. *Religion, Ritual and African Tradition: African Foundations.* Bloomington, IN: AuthorHouse, 2010.

Mbiti, John S. *Introduction to African Religion.* Oxford: Heinemann, 1975.

p. 11   China pulled ahead of the pack sewing up rare earth minerals throughout the African continent

Nesbit, Jeff. "China's Continuing Monopoly Over Rare Earth Minerals." *U.S. News and World Report,* April 2, 2013.

Jackson, Allison. "China Corners Rare Earths Market." *Technology Metals Research,* November 15, 2009.

French, Howard W. *China's Second Continent: How a Million Migrants Are Building a New Empire in Africa.* New York: Knopf, 2014.

Rice, Xan. "China's Economic Invasion of Africa." *The Guardian,* February 6, 2011.

p. 11   Even large retailers got in on the action, quickly locking in milk supplies and prices when rises in temperature were forecasted

Yano, Machiko, Hideyasu Shimadzu, and Toshiki Endo. "Modelling Temperature Effects on Milk Production: A Study on Holstein Cows at a Japanese Farm." *SpringerPlus* 3, no. 1. (2014). doi: 10.1186/2193-1801-3-129.

Broucek, Jan J., et al. "Effects of High Temperatures on Milk Production of Dairy Cows in East Central Europe." Sixth International Dairy Housing Conference Proceeding, Minneapolis, Minnesota, June 16–18, 2007. doi: 10.13031/2013.22792.

Mauger, Guillaume, Yoram Bauman, Tamilee Nennich, and Eric Salathé. "Impacts of Climate Change on Milk Production in the United States." *The Professional Geographer* 67, no. 1 (2015).

Griffin, Dan. "Hot Weather Affecting Milk Production on Wisconsin Farms." *WAOW.com.* July 18, 2013. http://www.waow.com/story/22880414/2013/07/ Thursday/hot-weather-affecting-milk-production-on-wisconsin-farms.

p. 11   Last year, Amazon filed for a patent on a blimp-like "floating warehouse"

Shead, Sam. "Amazon Is Considering Using Blimps as Huge Airborne Warehouses." *Business Insider*, December 29, 2016, http://www. businessinsider.com/amazon-blimps-airborne-warehouses-2016-12.

Brant, Tom. "Amazon Patents a Floating Drone Warehouse." *PC Magazine*, December 29, 2016. http://www.pcmag.com/ news/350670/amazon-patents-a-floating-drone-warehouse.

p. 13   Hospitals will use these same 3-D printers to produce prosthetic body parts on the fly

Khan, Amir. "How 3-D Printing Will Revolutionize Prosthetics." *U.S. News and World Report*, July 16, 2014. http://health.usnews. com/health-news/health-wellness/articles/2014/07/16/ how-3-d-printing-will-revolutionize-prosthetics.

Johnson, Steve. "3-D Printers to Make Human Body Parts? It's Happening." *The Mercury News.* January 28, 2015. http://www. mercurynews.com/2015/01/28/3d-printers-to-make-human-body-parts-its-happening/.

Starr, Michelle. "World's First 3D-Printed Apartment Building Constructed in China." *CNET.* January 19, 2015. https://www.cnet.com/newsworlds-first-3d-printed-apartment-building-constructed-in-china/.

Griffiths, Sarah. "Giant 3D Printer Creates 10 Full-Sized Houses in a DAY: Bungalows Built from Layers of Waste Materials Cost Less Than £3,000 Each." *Daily Mail*, April 28, 2014. http://www.dailymail. co.uk/sciencetech/article-2615076/Giant-3D-printer-creates-10-sized-houses-DAY-Bungalows-built-layers-waste-materials-cost-3-000-each.html.

Sapienza, James Derek. "Can We Really 3D Print Cars and Car Parts?" *Autos CheatSheet*, May 5, 2015. http://www.cheatsheet.com/automobiles/how-3-d-printing-will-transform-the-way-we-think-about-cars.html/?a=viewall.

Hornick, John. *3D Printing Will Rock the World.* North Charleston, SC: CreateSpace Independent Publishing Platform, 2015.

CHAPTER TWO
To See Or Not to See

p. 21   At one point there were more PhD's and engineers within thirty square miles than could be found in entire countries

A Half-Century of Learning: Historical Statistics on Educational Attainment in the United States, 1940 to 2000. Table 6. Percent of the Total Population 25 Years and Over with a Bachelor's Degree or Higher, by Sex, for the United States, Regions, and States: 1940 to 2000. United States Census Bureau, 2000.

Hoefler, Don. "Silicon Valley, USA." *Electronic News,* 1971.

Lécuyer, Christophe. *Making Silicon Valley: Innovation and the Growth of High Tech, 1930–1970.* Cambridge: MIT Press, 2006.

Paton, Luke, and Amy Bell. "50 U.S. Cities with the Most Doctoral Degree Holders." *Online PhD Programs.* http://www.online-phd-programs.org/50-u-s-cities-with-the-most-doctoral-degree-holders/.

Berlin, Leslie. "Why Silicon Valley Will Continue to Rule." *Backchannel.com.* May 1, 2015. https://backchannel.com/why-silicon-valley-will-continue-to-rule-cocbb441e22f.

p. 22   The world's first computer-aided design (CAD) system for electronic circuit design was born

Weisberg, David E. *The Engineering Design Revolution: The People, Companies and Computer Systems That Changed Forever the Practice of Engineering.* Cadhistory.net: 2008.

Bozdoc, Marian. *The History of CAD*. iMB: 2003. http://www.
   mbdesign.net/mbinfo/CAD-History.htm.

**p. 25   One of the primary barometers lenders use to qualify bor-
rowers is a mysterious algorithm called a FICO score**

"How FICO Became 'the' Credit Score." *bankrate.com*. December
   4, 2009.

Hill, Adriene. "A Brief History of the Credit Score." *Marketplace*,
   April 22, 2014. https://www.marketplace.org/2014/04/22/
   your-money/live-stage/brief-history-credit-score.

"What is a FICO score?" *MyFICO.com*. November 11, 2016. http://www.
   myfico.com/credit-education/credit-report-credit-score-articles/.

CHAPTER THREE
Jumping the Jar

**p. 31   In 2011, the popular American quiz show *Jeopardy!* laid
down the gauntlet**

Best, Jo. "IBM Watson: The Inside Story of How the Jeopardy-Winning
   Supercomputer Was Born, and What It Wants to Do Next."
   TechRepublic. September 10, 2013. http://www.techrepublic.com/
   article/ibm-watson-the-inside-story-of-how-the-jeopardy-winning-
   supercomputer-was-born-and-what-it-wants-to-do-next/.

Loftus, Jack "IBM Prepping 'Watson' Computer to Compete on
   Jeopardy!" *Gizmodo*, April 26, 2009.
   http://gizmodo.com/5228887/
   ibm-prepping-watson-computer-to-compete-on-jeopardy.

Baker, Stephen. *Final Jeopardy: The Story of Watson, the Computer That
   Will Transform Our World*. Boston: Mariner Books, 2012.

Kelly, John E. III, and Steve Hamm. *Smart Machines: IBM's Watson and
   the Era of Cognitive Computing*. New York: Columbia University
   Press, 2013.

p. 31   So, for example, if "The capital of the Ottoman Empire" appeared

"Dave Ferrucci at Computer History Museum: How It All Began and
What's Next." *IBM.com.* December 1, 2011. https://www.ibm.com/
blogs/research/2011/12/dave-ferrucci-at-computer-history-
museum-how-it-all-began-and-whats-next/

Loftus. "IBM Prepping 'Watson' Computer to Compete on Jeopardy!"

Best. "IBM Watson: The Inside Story of How the Jeopardy-Winning
Supercomputer Was Born, and What It Wants to Do Next."

Gondek, David. "How Watson 'Sees,' 'Hears,' and 'Speaks'
to Play Jeopardy!" *IBM Research* (blog). January 10, 2011.
https://www.ibm.com/blogs/research/2011/01/how-watson-sees-
hears-and-speaks-to-play-jeopardy/.

**p. 32   Enter Dr. David Ferrucci, Senior Manager of IBM's
Semantic Analysis and Integration Department**

"Dave Ferrucci at Computer History Museum: How It All Began
and What's Next."

Robinson, Gary (Big Data Technical Marketing, IBM Corporation)
in discussion with the author, 2015.

Best. "IBM Watson: The Inside Story of How the Jeopardy-Winning
Supercomputer Was Born, and What It Wants to Do Next."

Baker. *Final Jeopardy: The Story of Watson, the Computer That Will
Transform Our World.*

Greenemeier, Larry. "Will IBM's Watson Usher in a New Era of
Cognitive Computing?" *Scientific American*, November 13, 2013.

Kelly and Hamm. *Smart Machines: IBM's Watson and the Era of
Cognitive Computing.*

**p. 32   By the time the much-anticipated televised match of man
against machine rolled around**

Jackson, Joab. "IBM Watson Vanquishes Human Jeopardy Foes." *PC
World*, February 17, 2011. http://www.pcworld.com/article/219893/
ibm_watson_vanquishes_human_jeopardy_foes.html.

"Dave Ferrucci at Computer History Museum: How It All Began and What's Next."

Loftus. "IBM Prepping 'Watson' Computer to Compete on Jeopardy!"

Best. "IBM Watson: The Inside Story of How the Jeopardy-Winning Supercomputer Was Born, and What It Wants to Do Next."

p. 32   Studies reveal the human brain has a storage capacity of around 2.5 Petabytes

Reber, Paul. "What Is the Memory Capacity of the Human Brain?" *Scientific American*, May 1, 2010.

p. 33   On February 14, 2011, the first historic match featuring Watson and the top winners of *Jeopardy!* aired

"Dave Ferrucci at Computer History Museum: How It All Began and What's Next."

Loftus. "IBM Prepping 'Watson' Computer to Compete on Jeopardy!"

Jackson. "IBM Watson Vanquishes Human Jeopardy Foes."

Zimmer, Ben. "Is It Time to Welcome Our New Computer Overlords?" *The Atlantic*, February 17, 2011. https://www.theatlantic.com/technology/archive/2011/02/is-it-time-to-welcome-our-new-computer-overlords/71388/.

Raz, Guy. "Can a Computer Become a Jeopardy! Champ?" *All Things Considered*. National Public Radio, February 18, 2011. http://www.npr.org/2011/01/08/132769575/Can-A-Computer-Become-A-Jeopardy-Champ.

Best. "IBM Watson: The Inside Story of How the Jeopardy-Winning Supercomputer Was Born, and What It Wants to Do Next."

p. 34   We were no longer searching for a needle in a haystack

Laney, Doug. "Deja VVVu: Others Claiming Gartner's Construct for Big Data." *Gartner Blog Network*. January 14, 2012. http://blogs.gartner.com/doug-laney/deja-vvvue-others-claiming-gartners-volume-velocity-variety-construct-for-big-data/.

p. 35   Eric Schmidt, the Executive Chairman of Google explains the challenge volume and velocity presents

Vance, Jeff. "Big Data Analytics Overview." *Datamation.* June 25, 2013. http://www.datamation.com/applications/big-data-analytics-overview.html.

p. 35   Did you know that every minute of the day, two to three days' worth of new video content is uploaded to YouTube?

Siegler, M.G. "Every Minute, Just About a Day's Worth of Video is Now Uploaded to YouTube." *TechCrunch.* May 20, 2009. https://techcrunch.com/2009/05/20/every-minute-just-about-a-days-worth-of-video-is-uploaded-to-youtube/.

p. 35   Internet statistics site Pingdom claims we now send 144 billion emails a day

Outlook.com. "Did You Know 144.8 Billion Emails Are Sent Every Day?" *Mashable.* November 27, 2012. http://mashable.com/2012/11/27/email-stats-infographic/#aM9GvPUrwGqE.

p. 36   Recent studies showed that physicians would have to spend 160 hours a week reading medical journals to remain current in their field

Robinson, Gary (Big Data Technical Marketing, IBM Corporation) in discussion with the author, 2015.

Meier, Justin, Carla Remulla, Blanca Villanueva, and Emeline Wu. "Ethical Issues of Watson," *Watson in Healthcare.* http://watsonsmedicalcareer.weebly.com/ethical-issues.html

Mason, Barry. "Putting IBM Watson to Work in Healthcare." 2012. http://www.umsl.edu/~sauterv/DSS/Watson%20for%20Healthcare%20UMSL%20042512.pdf

"Watson Beats Doctors at Cancer Diagnosis." *QMED.* Accessed May 9, 2017. http://www.qmed.com/mpmn/gallery/image/watson-beats-doctors-cancer-diagnosis.

"Survey: How Doctors Read and What it Means To Patients."
    BusinessWire. July 22, 2014. http://www.businesswire.com/news/
    home/20140722005535/en/Survey-Doctors-Read-Means-Patients.

**p. 36   That same study showed that 81 percent of physicians
admitted spending five hours or less per month reviewing research**

Robinson, Gary (Big Data Technical Marketing, IBM Corporation)
    in discussion with the author, 2015.

Meier, Justin, Carla Remulla, Blanca Villanueva, and Emeline Wu.
    "Ethical Issues of Watson," *http://watsonsmedicalcareer.weebly.com/
    ethical-issues.html*

Mason. "Putting IBM Watson to Work in Healthcare."

"Survey: How Doctors Read and What it Means To Patients."

**p. 36   Dr. Steven Shapiro, Chief Medical and Science Officer at
University of Pittsburgh Medical Center**

Robinson, Gary (Big Data Technical Marketing, IBM Corporation)
    in discussion with the author, 2015.

Meier, Justin, Carla Remulla, Blanca Villanueva, and Emeline Wu.
    "Ethical Issues of Watson," *Watson in Healthcare*. http://watsons-
    medicalcareer.weebly.com/about.html

Mason. "Putting IBM Watson to Work in Healthcare."

**p. 37   They began with the Memorial Sloan Kettering Cancer
Center and WellPoint's oncology group**

Robinson, Gary (Big Data Technical Marketing, IBM Corporation)
    in discussion with the author, 2015.

Mathews, Anna Wilde. "Wellpoint's New Hire. What Is Watson?"
    The Wall Street Journal., September 12, 2011.

Best. "IBM Watson: The Inside Story of How the Jeopardy-Winning
    Supercomputer Was Born, and What It Wants to Do Next."

Upbin, Bruce. "IBM's Watson Gets Its First Piece of Business in
    Healthcare." *Forbes*. February 8, 2013. https://www.forbes.com/

sites/bruceupbin/2013/02/08/ibms-watson-gets-its-first-piece-of-business-in-healthcare/#1603851c5402.

p. 37   According to Dr. Samuel Nessbaum of WellPoint, Big Data has revolutionized oncology

IBM. "The Computing System that Won 'Jeopardy!' Is Helping Doctors Fight Cancer." *Business Insider*. February 4, 2015. http://www.businessinsider.com/sc/ibm-watson-and-medicine-2015-2.

"Watson Beats Doctors at Cancer Diagnosis."

"Doctors Call on Supercomputer Watson to Help Fight Cancer." *Reuters*, February 8, 2013.

p. 38   By 2013, revenues from Big Data jumped 58 percent climbing to $19 billion

Kelly, Jeff. "Big Data Vendor Revenue and Market Forecast 2013–2017." *Wikibon*. February 12, 2014. http://wikibon.org/wiki/v/Big_Data_Vendor_Revenue_and_Market_Forecast_2013-2017.

Jackson, Joab. "IBM Bets Big on Watson-Branded Cognitive Computing." *PCWorld*. January 9, 2014. http://www.pcworld.com/article/2086520/ibm-bets-big-on-watsonbranded-cognitive-computing.html.

p. 38   Then in 2009, Swedish-American start-up Recorded Future set out to achieve the impossible

Temple-Raston, Dina. "Predicting the Future: Fantasy or a Good Algorithm?" *Morning Edition*. National Public Radio, October 8, 2012.

Siegel, Eric. *Predictive Analytics: The Power to Predict Who Will Click, Buy, Lie, or Die*. Hoboken, NJ: Wiley, 2016.

Recordedfuture.com

p. 39   And in 2010, Recorded Future hit pay dirt. They predicted Yemen was headed for upheaval

Temple-Raston. "Predicting the Future: Fantasy or a Good Algorithm?

p. 39   Co-founder Christopher Ahlberg was elated

Temple-Raston. "Predicting the Future: Fantasy or a Good Algorithm?

p. 40   As news spread that 60 percent of drug overdoses in the U.S. were attributable to prescription opioids

Marr, Bernard. "How Big Data Helps to Tackle the No 1 Cause of Accidental Death in the US." *Forbes*, January 16, 2017.

Temple-Raston. "Predicting the Future: Fantasy or a Good Algorithm?"

p. 40   Using a patient's "past medical and pharmacy utilization, location, and demographic data"

Fuzzy Logix. "Fuzzy Logix Presents on Big Data Analytics and the Fight Against Opioid Addiction." Presentation at the Teradata Partners Conference, Atlanta, GA, September 10–15, 2016.

Temple-Raston. "Predicting the Future: Fantasy or a Good Algorithm?"

p. 40   Fuzzy Logix offers more than seven hundred predictive algorithms

"Fighting diabetes with data." *Fuzzylogix.com*. 2016. http://www. fuzzylogix.com/solutions/chronic-illness-predictive-modelling/.

p. 41   After twelve thousand hours of laboriously studying high-performing companies

"PearlHPS' PearlPredict Predicts Business Outcomes 12 Months Out." *Marketwired.com*. June 21, 2016. http://www.marketwired. com/press-release/pearlhpss-pearlpredict-predicts-business-outcomes-12-months-out-2136167.htm

p. 41   And, following several lean years—and self-doubt as to whether such a formula was even possible

"PearlHPS' PearlPredict Predicts Business Outcomes 12 Months Out."

p. 42   Using PearlHPS' Execution Analytics to compare and contrast

"PearlHPS' PearlPredict Predicts Business Outcomes 12 Months Out."

p. 42   We know which twelve-year-olds are likely to become binge drinkers

Singh, Maanvi. "Can We Predict Which Teens Are Likely to Binge Drink? Maybe." National Public Radio. July 2, 2014.

"Tests Can Predict Teens Most Likely to Binge Drink." *NHS Choices.* July 3. 2014. http://www.nhs.uk/news/2014/07July/P. s/Tests-can-predict-teens-most-likely-to-binge-drink.aspx.

Crutzen, Rik, Philippe J. Giabbanelli, Astrid Jander, Liesbeth Mercken, and Hein de Vries. "Identifying Binge Drinkers Based on Parenting Dimensions and Alcohol-Specific Parenting Practices: Building Classifiers on Adolescent-Parent Paired Data." *BMC Public Health* 15, no. 747 (2015). DOI: 10.1186/s12889-015-2092-8.

Mazumdar, Agneeth and X. J. Selman. "5 Bizarre Ways Brain Scans Can Predict the Future." *Cracked.com.* October 25, 2012. http://www.cracked.com/article_20073_5-bizarre-ways-brain-scans-can-predict-future.html.

CHAPTER FOUR
Unintended, But Not Unanticipated

p. 47   Claire Andre and Manuel Velasquez, professors in Business Administration at Santa Clara University, were among the first to sound the alarm

Andre, Claire, and Manuel Velasquez. "Who Should Pay? The Product Liability Debate." Santa Clara University Markkula Center for Applied Ethics. November 20, 2015. https://www.scu.edu/ethics/focus-areas/business-ethics/resources/who-should-pay-the-product-liability-debate/.

Polinsky, A. Mitchell, and Steven Shavell. "The Uneasy Case for Product Liability." *Harvard Law Review* 123 (2010): 1436.

p. 48   LexisNexis Market Intelligence reported that in 1999, nearly five thousand personal injury product liability cases were filed in US Federal Court

LexisNexis.com

"Civil Cases." *Bureau of Justice Statistics.* Last revised on April 4, 2017. https://www.bjs.gov/index.cfm?ty=tp&tid=45

p. 48   Within five years that number doubled. Two years after that, caseloads doubled again

Stickel, Amy I. "Product Liability: A Trend Still Worth Watching." *LexisNexis Market Intelligence Report.* 2005.

p. 48   In a study by Jury Verdict Research, the median award for product liability cases jumped

Stickel, Amy I. "Product Liability: A Trend Still Worth Watching."

p. 49   Andre and Velasquez sum up the damage

Andre and Velasquez. "Who Should Pay? The Product Liability Debate."

p. 49   Blitz USA was one of the fatalities

Miller, Steve. "Blame the Lawsuits for Shutting Down Blitz USA...Or Blame Blitz USA's Products for Causing the Lawsuits?" *Protect Consumer Justice.* August 14, 2012. http://www.protectconsumerjustice.org/blame-the-lawsuits-for-shutting-down-blitz-usa-or-blame-blitz-usas-products-for-causing-the-lawsuits.html.

Krauss, Clifford. "A Factory's Closing Focuses Attention on Tort Reform." *New York Times*, October 4, 2012. http://www.nytimes.com/2012/10/05/business/in-a-shuttered-gasoline-can-factory-the-two-sides-of-product-liability.html.

p. 49   But, according to Blitz CEO Rocky Flick, shields would not have prevented the explosions that occurred

Krauss. "A Factory's Closing Focuses Attention on Tort Reform."

p. 50   Emory University School of Law professor Frank Vandall takes a pragmatic view

Krauss. "A Factory's Closing Focuses Attention on Tort Reform."

p. 50   The company sold more than fourteen million containers, and experienced fewer than two problems for every million sold

Krauss. "A Factory's Closing Focuses Attention on Tort Reform."

p. 50 $30 million in legal fees, an additional $30 plus million in insurance company settlements

Krauss. "A Factory's Closing Focuses Attention on Tort Reform."

p. 50   One of the most egregious, best known liability cases occurred in 1994

Liebeck v. McDonald's Restaurants, P.T.S., Inc., Bernalillo County, N.M. Dist. Ct. (1994).

Gerlin, Andrea. "A Matter of Degree: How a Jury Decided that a Coffee Spill Is Worth $2.9 Million." *Wall Street Journal*, September 1, 1994.

Burtka, Allison Torres. "Liebeck v. McDonald's." *American Museum of Tort Law*. June 13, 2016. https://www.tortmuseum.org/liebeck-v-mcdonalds/.

p. 51   After a lengthy trial, the jury determined that McDonald's was 80 percent responsible

Liebeck v. McDonald's Restaurants, P.T.S., Inc.

Gerlin. "A Matter of Degree: How a Jury Decided that a Coffee Spill Is Worth $2.9 Million."

Burtka. "Liebeck v. McDonald's."

p. 51   In 2009, a suit was brought against the manufacturers of Bluetooth headsets

Ogg, Erica. "Motorola Sued Over Potential Bluetooth Hearing Loss." *CNET*.

October 30, 2006. https://www.cnet.com/news/motorola-sued-over-potential-bluetooth-hearing-loss/.

Glenn, Brandon. "Motorola Sued Over Bluetooth Headsets." *Crain's Chicago Business.* October 18, 2006. http://www.chicagobusiness.com/article/20061018/NEWS04/200022502/motorola-sued-over-bluetooth-headsets.

p. 51   Then there is the story of twenty-seven-year-old Daniel Dukes, who, in 1999, deliberately hid inside Orlando's Sea World until it closed

Associated Press. "Park Is Sued Over Death of Man in Whale Tank." *New York Times,* September 21, 1999.

p. 51   In a humorous article, journalists Brett Nelson and Katy Finneran

Nelson, Brett, and Katy Finneran. "Dumbest Warning Labels." *Forbes,* February 23, 2011.

Dorigo Jones, Bob. *Remove Child Before Folding: The 101 Stupidest, Silliest, and Wackiest Warning Labels Ever.* New York: Warner Books, 2007.

Green, Joey, Tony Dierckins, and Tim Nyberg. *The Warning Label Book.* New York: St. Martin's Griffin, 1998.

Koon, Jeff, and Andy Powell. *Wearing of This Garment Does Not Enable You to Fly: 101 Real Dumb Warning Labels.* New York: Free Press, 2008.

p. 53   In 2012, on an unseasonably warm Spring day, the body of American football star Junior Seau

Moore, David Leon, and Erik Brady. "Junior Seau's Final Days Plagued by Sleepless Nights." *USA Today,* June 2, 2012.

Greene, David. "Seau's Suicide Helped to Make Concussions in Football a National Issue." *Morning Edition,* National Public Radio. December 22, 2015.

Trotter, Jim. *Junior Seau: The Life and Death of a Football Icon.* New York: Houghton Mifflin Harcourt, 2015.

Farmer, Sam. "Junior Seau Had Brain Disease When He Committed Suicide." *Los Angeles Times*, January 10, 2013.

p. 53   In 2006, defensive lineman Shane Dronett began complaining of episodes of confusion, rage, and paranoia

Smith, Stephanie. "Ex-Falcons Lineman Had Brain Disease Linked to Concussions." *CNN.* April 1, 2011. http://www.cnn.com/2011/HEALTH/04/01/brain.concussion.dronett/.

p. 53   Pro-Bowlers Dave Duerson and Andre Waters met similar ends. As did young Penn State lineman Owen Thomas

Solotaroff, Paul. "Dave Duerson: The Ferocious Life and Tragic Death of a Super Bowl Star." *Mens Journal*, May 2, 2011.

Farrey, Tom. "Pathologist Says Waters' Brain Tissue Had Deteriorated." *ESPN.com.* January 19, 2007. http://www.espn.com/nfl/news/story?id=2734941.

Park, Madison. "College Football Player Who Committed Suicide Had Brain Injury." *CNN.com.* September 14, 2010. http://www.cnn.com/2010/HEALTH/09/14/thomas.football.brain/.

p. 54   CTE occurs when a protein called tau builds in the brain after repeated blows to the head

Salvitti, Anthony. *Chronic Traumatic Encephalopathy."* CreateSpace Independent Publishing Platform, 2013.

Fainaru-Wada, Mark, and Steve Fainaru. *League of Denial: The NFL, Concussions, and the Battle for Truth.* New York: Three Rivers Press, 2014.

p. 54   To date, more than five thousand players have sued the NFL over CTE

Perez, A.J. "NFL Faces New Concussion-Related Lawsuit." *USA Today*, March 29, 2016.

Omalu, Bennet. *A Historical Foundation of CTE in Football Players: Before the NFL, There was CTE*. Bennet Omalu, MD, 2014.

Salvitti. *Chronic Traumatic Encephalopathy*." CreateSpace Independent Publishing Platform. November 10, 2013.

Fainaru-Wada and Fainaru. *League of Denial: The NFL, Concussions, and the Battle for Truth*.

Almasy, Steve, and Jill Martin. "Judge Approves NFL Concussion Lawsuit Settlement." *CNN.com*. April 22, 2015. http://www.cnn.com/2015/04/22/us/nfl-concussion-lawsuit-settlement/.

Hayes, Ashley, and Michael Martinez. "Former NFL Players: League Concealed Concussion Risks." *CNN.com*. July 20, 2011. http://www.cnn.com/2011/HEALTH/07/20/nfl.lawsuit.concussions/.

p. 55   So much so that the Boston University School of Medicine CTE Center reported that the early signs of CTE have been spotted in fourteen out of fifteen former NFL players studied

"Boston University Researchers Report NHL Player Derek Boogaard Had Evidence of Early Chronic Encephalopathy." *Boston University School of Medicine*. December 6, 2011. https://www.bumc.bu.edu/busm/2011/12/06/boston-university-researchers-report-nhl-player-derek-boogaard-had-evidence-of-early-chronic-encephalopathy/.

Abreu, Marcos A., Fred J. Cromartie, Brandon D. Spradley, and the United States Sports Academy. "Chronic Traumatic Encephalopathy (CTE) and Former National Football League Player Suicides." *The Sport Journal*, January 29, 2016.

Hayes and Martinez. "Former NFL Players: League Concealed Concussion Risks."

p. 56   Here, journalist Jonathan Tamari cuts to the chase

Tamari, Jonathan. "Seau and the Looming Cloud Over the NFL." *Philly.com*. May 3, 2012. http://www.philly.com/philly/blogs/inq-eagles/Seau-and-the-looming-cloud-over-the-NFL.html.

p. 56   The NFL is a $75 billion franchise

Gaines, Cork. "The 32 NFL Teams Are Worth as Much as Every MLB
    and NBA Team Combined." *Business Insider*. September, 15, 2016.
    http://www.businessinsider.com/nfl-teams-value-2016-9.

**p. 57   In 2015 a federal judge approved a $1 billion settlement
by the NFL**

Almasy and Martin. "Judge Approves NFL Concussion Lawsuit
    Settlement."

Perez. "NFL Faces New Concussion-Related Lawsuit."

p. 57   And though the NFL's official position had previously been

Martin, Jill. "NFL Acknowledges CTE Link with Football.
    Now What?" *CNN.com*. March 16, 2016. http://www.cnn.
    com/2016/03/15/health/nfl-cte-link/.

Farmer, Sam. "NFL Executive's Comments on CTE Quickly Attract
    Notice." *Los Angeles Times*, March 15, 2016.

**p. 59   Then at 9:20 p.m., tragedy struck. Terrorists opened fire at
six popular gathering places, killing 130 and injuring 368**

Steafel, Eleanor, Rory Mullholland, Rozina Sabur, Edward Malnick,
    Andrew Trotman, and Nicola Harley. "Paris Terror Attack:
    Everything We Know on Wednesday Evening." *The Telegraph*,
    November 18, 2015.

Tranton, Phillip. *Paris Attacks: Isis Spreading into Europe*. CreateSpace
    Independent Publishing Platform: 2015.

de la Hamaide, Sybille. "Timeline of Paris Attacks According to Public
    Prosecutor." *Reuters*. November 14, 2015. http://www.reuters.com/
    article/us-france-shooting-timeline-idUSKCN0T31BS20151114.

p. 59   It was the worst attack on French soil since World War II

"Paris Attacks: Hollande Blames Islamic State for 'Act of War.'" *BBC
    News*, November 14, 2015.

Nossiter, Adam, Aurelien Breeden, and Katrin Bennhold. "Three
    Teams of Coordinated Attackers Carried Out Assault on Paris,

Officials Say: Hollande Blames ISIS." *New York Times*, November 14, 2015.

Heneghan, Tom. "Hollande Says Paris Attacks 'an act of war' by Islamic State." *Reuters*, November 14, 2015.

p. 59   The police raided two hundred locations

Associated Press. "Paris Attacks: Identification and Arrests Start Piling Up." *CBS News*. November 15, 2015. http://www.cbsnews.com/news/paris-attacks-identifications-and-arrests-start-piling-up/.

Mullen, Jethro, Don Melvin, and Paul Armstrong. "Terror in Paris: What We Know So Far." *CNN.com*. November 15, 2015. http://www.cnn.com/2015/11/13/europe/paris-attacks-at-a-glance/.

"Paris Attacks: France Mobilises 115,000 Security Personnel." *BBC News*. November 17, 2015. http://www.bbc.com/news/world-europe-34840699.

de la Hamaide. "Timeline of Paris Attacks According to Public Prosecutor."

Burke, Jason. "Brussels to Stay in Lockdown 'For at Least Another Week.'" *The Guardian*, November 23, 2015.

Tranton. *Paris Attacks: Isis Spreading into Europe.*

p. 60   For example, authorities already had a lengthy dossier on Ismaël Omar Mostefaï

Farmer, Ben, Victoria Ward, Gordon Rayner, Dabid Barrett, Patrick Sawer, Luke Heighton, Camilla Turner, Rory Mullholland, Mathew Holehouse, David Chazan, Henry Samuel, Lexi Finnigan, Eleanor Steafel, Isabelle Fraser, and James Rothwell. "Who Is Salah Abdeslam and Who Were the Paris Terrorists? Everything We Know about ISIL Attackers." *The Telegraph*, March 18, 2016.

Associated Press. "Paris Attacks: Identification and Arrests Start Piling Up."

Callus, Andrew, and Bate Felix. "French Police Identify One of the Assailants in Paris Attacks." *Reuters*, November 15, 2015.

p. 61 In a chilling 2014 video he posted on the Internet

Dahlburg, John-Thor, Lori Hinnant, and Jamey Keaten. "The Mastermind Behind the Paris Attacks." *Associated Press*, November 18, 2015.

p. 62 Just about the time I was wondering why the police didn't piece together this evidence earlier

Abduhl-Zahra, Qassim. "Iraq Warned of Attacks Before Paris Assault." *Associated Press*, November 15, 2015.

Hinnant, Lori, Thomas Adamson. "France Bombs Islamic State HQ, Hunts Attacker Who Got Away." *Associated Press*, November 15, 2015.

p. 63 By the end of 2015, nearly four hundred thousand Syrian refugees were on the move in Europe

Myre, Greg. "The Migrant Crisis, By the Numbers." *Parallels*, National Public Radio. September 8, 2015.

p. 63 In 2004, ten bombs in backpacks were placed on four commuter trains in Madrid

"Spanish Indictment on the Investigation of 11 March." *El Mundo*, February 16, 2010.

p. 63 In 2011, thirty-two-year-old Anders Behring Breivik detonated a fertilizer bomb

Associated Press. "Norway Massacre Gunman Anders Breivik Declared Sane, Gets 21-Year Sentence." *NBC*, August 24, 2012.

p. 66 In 1990, HBO released a disturbing documentary titled Child of Rage

"Child of Rage." TV short. Produced by Anne H. Cohn, Dalton Delan, and Gaby Monet. Home Box Office. 1990.

Magid, Ken. *High Risk: Children Without a Conscience*. New York: Bantam Books, 1987.

Patel, Leela. *A Sociopath Is Born*. CreateSpace Independent Publishing Platform, 2016

p. 67   In 1984, Joshua Phillips' mother was cleaning his room

Alphonse, Lylah M. "Can a Kid Be a Psychopath?" *Yahoo*. May 14, 2012.
   https://ca.style.yahoo.com/blogs/parenting/kid-psychopath-
   221400341.html.

p. 68   In 1993, Jon Venables and Robert Thompson

Alphonse. "Can a Kid Be a Psychopath?"

p. 68   In 2009, fifteen-year-old Alyssa Bustamante confessed

Alphonse. "Can a Kid Be a Psychopath?"

p. 68   Dr. Paul Frick, a psychiatrist specializing in child psychopathy at the University of New Orleans

Kahn, Jennifer. "Can You Call a 9-Year-Old a Psychopath?" *New York
   Times*, May 11, 2012.

p. 69   In her recent article "Can a Kid be a Psychopath?"

Alphonse. "Can a Kid be a Psychopath?"

p. 70   In 2003, thirteen years after the US Office of Biological and Environmental Research launched the Human Genome Project

Nova: Cracking the Code of Life. Directed by Elizabeth Arledge.
   Documentary. New York: Public Broadcasting System, 2001.

Chial, Heidi. "DNA Sequencing Technologies Key to the Human Genome
   Project." *Scitable*. January 25, 2016. https://www.nature.com/scitable/
   topicp. /dna-sequencing-technologies-key-to-the-human-828

Venter, J. Craig. *A Life Decoded: My Genome: My Life*. New York:
   Penguin, 2007.

Yount, Lisa. *Craig Venter: Dissecting the Genome*. New York: Chelsea
   House Publishers, 2011.

Shreeve, James. *The Genome War: How Craig Venter Tried to Capture the
   Code of Life and Save the World*. New York: Ballantine Books, 2004.

p. 70   From sickle cell anemia to cystic fibrosis to muscular dystrophy to Huntington's disease and many cancers

"The Age of the Red Pen." *The Economist*, August 22, 2015.

Burney, Tabinda J., and Jane C Davies. "Gene Therapy for the Treatment of Cystic Fibrosis." *The Application of Clinical Genetics* 5 (2012): 29–36. doi:10.2147/TACG.S8873.

Smith, Moyra. *Seeking Cures: Design of Therapies for Genetically Determined Diseases.* Oxford: Oxford University Press, 2013.

Panno, Joseph. *Gene Therapy: Treatments and Cures for Genetic Diseases.* New York: Facts on File, 2010.

Weintraub, Karen. "Genetic Treatments for Sickle Cell." *Scientific American.* May 1, 2016. https://www.scientificamerican.com/article/genetic-treatments-for-sickle-cell/.

Chamberlain, Jeffrey S. "Gene therapy of muscular dystrophy." *Human Molecular Genetics* 11, no. 20 (2002): 2355–2362.

p. 71   Author of the bestselling book *The Zone*

Sears, Barry. *Enter the Zone: A Dietary Road Map.* New York: Regan Books, 1995.

p. 74   The following day an article in the *Los Angeles Times* suggested that those who aren't successful may be genetically predisposed to failure

Sapolsky, Robert M. "Rich Brain, Poor Brain." *Los Angeles Times,* October 18, 2013.

CHAPTER FIVE
Predaptation

p. 82   The financial institution we know today as Union Bank began 153 years ago

Loomis, Noel M. *Wells Fargo.* New York: Clarkson N. Potter, Inc., 1968.

Cronise, Titis Fey. *The Natural Wealth of California.* San Francisco: H.H. Bancroft & Company, 1868.

Maines, Penny. *Union Bank: 150 Years of History*. Virginia Beach, VA:
    The Donning Company Publishers, 2014.

**p. 84   What occurred next is largely known through Harpending's
journals**

Union Bank. *Union Bank: 150 Years of History*.
www.californiahistoricalsociety.org.

p. 86   In the 1800's, economist Alfred Marshall shocked the world

Marshall, Alfred. *Principles of Economics*. Unabridged 8th ed. New
    York: Cosimo Classics, 2009.

p. 87   Soon other social scientists followed Marshall's lead

Schumpeter, Joseph A. *Capitalism, Socialism, and Democracy*. 3rd ed.
    New York: Harper Perennial Modern Thought, 2008.

**p. 87   But Jefferson, an early proponent of democracy, argued
that even a foundational doctrine such as the US Constitution must
accommodate change**

Jefferson, Thomas. Letter from Thomas Jefferson to Adamantios
    Coray, October 31, 1823. *Founders Online*. https://founders.archives.
    gov/documents/Jefferson/98-01-02-3837.

**p. 88   And since 1791, the Constitution has been amended twenty-
seven times**

"Measures Proposed to Amend the Constitution." *United States
    Senate*. Accessed May 10, 2017. https://www.senate.gov/reference/
    measures_proposed_to_amend_constitution.htm.

**p. 89   Akio Morita's first product—a simple rice cooker—burned
so much rice he sold fewer than one hundred units**

"50 Famously Successful People Who Failed At First." *OnlineCollege.
    org*. February 16, 2010. http://www.onlinecollege.org/2010/02/16/50-
    famously-successful-people-who-failed-at-first/.

p. 90   Bill Gates' first venture, Traf-O-Data, went down in flames

"50 Famously Successful People Who Failed At First."

p. 90   Dan Brown's first two books received little attention until he penned *The Da Vinci Code*

Langley, William. "Dan Brown: A Success Story Even More Implausible Than His Plots." *The Telegraph*, April 25, 2009.

"Who Dan Brown Was Before the Da Vinci Code." *Now Novel*. Accessed May 10, 2017. http://www.nownovel.com/blog/dan-brown-da-vinci-code/.

p. 90   And many people are shocked to learn that van Gogh sold only one painting while he was alive

Scott, Steve. "Top 10 Successful People Who Have Failed." Steve Scott Site (blog). Accessed May 10, 2017. http://www.stevescottsite.com/successful-people-who-have-failed.

Marder, Lisa. "The Lore: Van Gogh Sold Only One Painting During His Life." *ThoughtCo*. May 31, 2016. https://www.thoughtco.com/van-gogh-sold-only-one-painting-4050008.

p. 90   Dr. Jamer Hunt teaches social and cultural anthropology at The New School in New York City

Hunt, Jamer. "Among Six Types of Failure, Only a Few Help You Innovate." *Co.Design*. June 27, 2011. https://www.fastcodesign.com/1664360/among-six-types-of-failure-only-a-few-help-you-innovate.

p. 91   Hunt observes, "There's no question that out of failure—even 'abject failure'—we emerge transformed"

Hunt,. "Among Six Types of Failure, Only a Few Help You Innovate."

p. 93   Vice President of Human Resources Kent Hansen observed

Hansen, Kent (Dole Fresh Fruits and Vegetables) in discussion with the author, January 11, 2015.

p. 98   As of today, scientists have classified more than 12,500 species of ants

Ward, Philip S. "Phylogeny, Classification, and Species-Level Taxonomy of Ants (Hymenoptera: Formicidae)." *Zootaxa* 1668

(2007): 549–563. http://www.mapress.com/zootaxa/2007f/zt01668p563.pdf.

Johnson, Norman. "Hymenoptera Name Server. Formicidae Species Count." Ohio State University. 2007. http://atbi.biosci.ohio-state.edu/hymenoptera/tsa.sppcount?the_taxon=Formicidae.

p. 99　As of 2017, there were an estimated one trillion euros in circulation

"Banknotes and Coins Circulation." *European Central Bank.* Accessed March 14, 2017. https://www.ecb.europa.eu/stats/policy_and_exchange_rates/banknotes+coins/circulation/html/index.en.html.

p. 101　Having no other choice, the International Monetary Fund, Eurogroup, and European Central Bank jumped

"Europe and IMF Agree €110 Billion Financing Plan With Greece." *International Monetary Fund.* May 2, 2010. http://www.imf.org/external/pubs/ft/survey/so/2010/car050210a.htm.

Ewing, Jack, and Liz Alderman. "Bailout Money Goes to Greece, Only to Flow Out Again." *New York Times,* July 30, 2015.

Castle, Stephen. "Europe Agrees on New Bailout to Help Greece Avoid Default." *New York Times,* February 20, 2012.

European Commission Directorate-General for Economic and Financial Affairs. "The Second Economic Adjustment Programme for Greece." *Occasional Papers* 159. July 29, 2013. http://ec.europa.eu/economy_finance/publications/occasional_paper/2013/pdf/ocp159_en.pdf.

Wearden, Graeme, and Nick Fletcher. "Eurozone Crisis Live: ECB to Launch Massive Cash Injection". *The Guardian,* February 29, 2012. https://www.theguardian.com/business/2012/feb/29/eurozone-debt-crisis-ecb-loans-ltro#block-16.

p. 102　Most of us have forgotten that in 1944, as World War II was coming to an end

Palairet, Michael R. *The Four Ends of the Greek Hyperinflation of 1941-1946.* Copenhagen: Museum Tusculanum Press, 2000.

"The Worst Cases of Hyperinflation Situations of All Time." *CNBC. com.* Accessed May 10, 2017. http://www.cnbc.com/2011/02/14/ The-Worst-Hyperinflation-Situations-of-All-Time.html?slide=3.

Siklos, Pierre L. *Great Inflations of the 20th Century: Theories, Policies and Evidence.* Cheltenham, England: Edward Elgar Publishing, 1995.

p. 102   The Viceroy doesn't come equipped with this defense

Ritland, David B. "Mimicry-Related Predation on Two Viceroy Butterfly (Limenitis archippus) Phenotypes." *The American Midland Naturalist* 140, no. 1 (1998): 1–20.

van Zandt Brower, Jane. "Experimental Studies of Mimicry in Some North American Butterflies: Part I. The Monarch, Danaus plexippus, and Viceroy, Limenitis archippus archippus." *Evolution* 12, no. 1 (1958): 32–47.

Platt, Austin P., Raymond P. Coppinger, and Lincoln P. Brower. "Demonstration of the Selective Advantage of Mimetic Limenitis Butterflies Presented to Caged Avian Predators." *Evolution* 25, no. 4 (1971): 692–701.

Ritland, David B. "Comparative Unpalatability of Mimetic Viceroy Butterflies (Limenitis archippus) from Four South-Eastern United States Populations." *Oecologia* 103, no. 3 (1995): 327–336.

p. 102   The syrphid fly adopted a similar strategy

Penney, Heather D., Christopher Hassall, Jeffrey H. Skevington, Brent Lamborn, and Thomas N. Sherratt. "The Relationship Between Morphological and Behavioral Mimicry in Hover Flies (Diptera: Syrphidae)." *The American Naturalist* 183, no. 2 (2013): 281–289.

Polidori, Carlo, Jose Nieves-Aldrey, Francis Gilbert, and Graham E. Rotheray. "Hidden in Taxonomy: Batesian Mimicry by a Syrphid Fly Towards a Patagonian Bumblebee." *Insect Conservation and Diversity* 7, no, 1 (2014): 32–40.

Hadley, Debbie. "What Is Batesian Mimicry?" *ThoughtCo.* January 5, 2017. https://www.thoughtco.com/what-is-batesian-mimicry-1968038.

p. 103   *Bloomberg News* **ranks the healthcare systems of Hong Kong and Singapore**

"Bloomberg Ranks the World's Most Efficient Health Care Systems." *Advisory Board.* August 28, 2013. https://www.advisory.com/daily-briefing/2013/08/28/bloomberg-ranks-the-worlds-most-efficient-health-care-systems.

p. 103   And according to the Worldwide Governance Indicators project

Hetter, Katia. "Where Are the World's Happiest Countries?" *CNN.com.* March 21, 2017. http://www.cnn.com/2017/03/20/travel/worlds-happiest-countries-united-nations-2017/.

"The 25 Best Governments in the World." *WorldAtlas.* Accessed May 10, 2017. http://www.worldatlas.com/articles/the-best-governments-in-the-world.html.

p. 109   At their height, Solyndra had one thousand systems in the field

Prend, David (Rockport Capital Partners), in discussion with the author, 2010.

"New Shape of Solar." *Solyndra,LLC.* 2008. http://www.solyndra.com/technology-products/cylindrical-module/

Biello, David. "Cylindrical Solar Cells Give a Whole New Meaning to Sunroof." *Scientific American,* October 7, 2008. https://www.scientificamerican.com/article/cylindrical-solar-cells-give-new-meaning-to-sunroof/.

Wang, Ucilia. "Solyndra Rolls Out Tube-Shaped Thin Film." *Greentech Media.* October 7, 2008. https://www.greentechmedia.com/articles/read/solyndra-rolls-out-tube-shaped-thin-film-1542.

Green, Hank. "Tubular Solar Panels Slash Costs, Boost Efficiency." *EcoGeek*. October 7, 2008. http://ecogeek.wpengine.com/?s= Tubular+Solar+Panels+Slash+Costs%2C+Boost+Efficiency.

Ucilia Wang. "Solyndra Works on 1M Sq. Ft. Project in SoCal." Greentech Media. July 16, 2009. https://www.greentechmedia. com/articles/read/solyndra-works-on-1m-sq.-ft.-project-in-socal.

Sidhu, Ikhlaq, Shomit Ghose, and Paul Nerger. "Solyndra 2011 Case Study." Fung Institute, UC Berkeley College of Engineering. January 2012. https://ikhlaqsidhu.files.wordpress.com/2012/09/ solyndra_case_v3-3-1kr.pdf.

p. 109   Given Solyndra's early success—and the solid track record of the founders

"Obama Administration Offers $535 Million Loan Guarantee to Solyndra, Inc." Energy.gov. March 20, 2009. https://energy. gov/articles/obama-administration-offers-535-million-loan-guarantee-solyndra-inc.

Sidhu, Ghose, and Nerger. "Solyndra 2011 Case Study."

p. 109   In 2008, the price of solar modules took a nose dive from $4.20 per watt to $1.20 per watt

Harder, Amy. "Deconstructing Solyndra." National Journal, November 23, 2011.

Sidhu, Ghose, and Nerger. "Solyndra 2011 Case Study."

Bielo, David. "How Solyndra's Failure Promises a Brighter Future for Solar Power." Scientific American, October 12, 2011.

Energy and Commerce Committee, Subcommittee on Oversight and Investigations, and U.S. House of Representatives. Solyndra and the Department of Energy Loan Guarantee Program: House Hearings on Stimulus Funding for Solar Energy Company. Progressive Management, 2011.

p. 111   In 1867, Karl Marx published *Das Kapital*

Marx, Karl. *Das Kapital*: A Critique of Political Economy. CreateSpace Independent Publishing Platform, 2011.

Mandel, Ernest. Marxist Economic Theory, Vols. 1 and 2. New York: Monthly Review Press, 1970.

Morishima, Michio. Marx's Economics: A Dual Theory of Worth and Growth. Cambridge: Cambridge University Press, 1973.

p. 111   Nor could he have imagined that China would corner 95 percent of the total rare earth minerals

Nesbit. "China's Continuing Monopoly Over Rare Earth Minerals."

Jackson. "China Corners Rare Earths Market."

French. China's Second Continent: How a Million Migrants Are Building a New Empire in Africa.

Rice. "China's Economic Invasion of Africa."

p. 112   The horned lizard shoots blood from its eyes

Ciprian, Steve. "15 Unusual Animal Defense Mechanisms." Yur Topic. June 20, 2013. http://www.yurtopic.com/science/nature/animal-defense-mechanisms.html.

Middendorf, George A. III, Wade C. Sherbrooke, and Eldon J. Braun. "Comparison of Blood Squirted from the Circumorbital Sinus and Systemic Blood in a Horned Lizard, Phrynosoma cornutum." The Southwestern Naturalist 46, no. 3(2001): 384–387.

Hodges, Wendy L. "Defensive Blood Squirting in Phrynosoma ditmarsi and a High Rate of Human-Induced Blood Squirting in Phrynosoma asio." The Southwestern Naturalist 49, no. 2 (2004): 267–270.

Wang, Bin, Wen Yang, Vincent R. Sherman, and Marc A. Meyers. "Pangolin Armor: Overlapping, Structure, and Mechanical Properties of the Keratinous Scales." Acta Biomaterialia 41, no. 6 (2016): 60–74. doi: 10.1016/j.actbio.2016.05.028.

Wilsdon, Christina. Animal Defenses. New York: Chelsea House
Publishers, 2009.

Riehecky, Janet. Camouflage and Mimicry: Animal Weapons and
Defenses. North Mankato, Minn.: Capstone Press, 2012.

Cormier, Zoe. "Termites Explode to Defend Their Colonies."
Nature, July 26, 2012.

**p. 115   During the fourteenth Century Japanese ninjas began
wearing black to camouflage themselves during night missions**

Grace F.M. "The History of Camouflage." Ezine Articles.
January 10, 2012. http://ezinearticles.com/?The-History-of-
Camouflage&id=6808569.

Szczepanski, Kallie. "The Ninja of Japan." ThoughtCo. April 14, 2017.
https://www.thoughtco.com/history-of-the-ninja-195811.

**p. 115   In the eighteenth century, German Jäger rifle units
switched to drab brown and green clothing to blend in with their
natural surroundings**

Grace F.M. "The History of Camouflage."

**p. 120   Convinced advertisers could not afford to stay away if his
numbers continued to climb, Limbaugh grew more emboldened**

Kurtzman, Daniel."Rush Limbaugh Quotes." ThoughtCo. July 30, 2016.
https://www.thoughtco.com/rush-limbaugh-quotes-2734660.

Epstein, Ethan. "Is Rush Limbaugh in Trouble?" POLITICO
Magazine, May 24, 2016.

**p. 121   Then in February 2012, the man responsible for changing
the face of AM radio called Georgetown University Law Student
Sandra Fluke**

Johnson, Jenna. "Georgetown President Defends Sandra Fluke,
Blasts Rush Limbaugh." The Washington Post, March 2, 2012.
https://www.washingtonpost.com/local/education/georgetown-
president-defends-student-blasts-limbaugh/2012/03/02/
gIQAnE2omR_story.html?utm_term=.182dc26dd714.

Bingham, Amy. "Sponsors Pull Ads from Rush Limbaugh's Radio Show Over 'Slut' Comments." ABC News. March 4, 2012. http://abcnews.go.com/blogs/politics/2012/03/advertisers-pull-funds-from-rush-limbaughs-radio-show-over-slut-comments/.

Tartar, Andre. "Rush Limbaugh's 'Slut' Attack Costs Him Advertisers." New York. March 4, 2012. http://nymag.com/daily/intelligencer/2012/03/rush-limbaughs-slut-attack-costs-him-advertisers.html.

Mirkinson, Jack. "Rush Limbaugh Advertisers Pull Commercials in Wake of Sandra Fluke 'Slut' Firestorm." Huffington Post. March 2, 2012. http://www.huffingtonpost.com/2012/03/02/rush-limbaugh-sleep-train-sandra-fluke-slut_n_1315900.html.

p. 124   They discovered the dung beetle

Dacke, Marie, Emily Baird, Marcus Byrne, Clarke H. Scholtz, and Eric J. Warrant. "Dung Beetles Use the Milky Way for Orientation." Current Biology 23, no. 4 (2013): 298–300. DOI: 10.1016/j.cub.2012.12.034.

"Dung Beetles Navigate by the Stars." The Guardian, January 25. 2013.

p. 125   In 2015, chairman of the World Conference on Disaster Management Paul Kovacs

Kovacs, Paul. Presentation at the 2015 Conference on Disaster Management, Toronto, Canada, June 8–11, 2015.

p. 128   When the number of tigers remaining on Earth reached 3,200

Langlin, Katie. "Pictures: The World's Tigers—There Are Only 3200 Left in the Wild." National Geographic. July 29, 2015. http://news.nationalgeographic.com/news/2014/07/pictures/140729-tigers-conservation-cubs-hunting-environment-science/.

Chappell,Bill. "Number Of Wild Tigers Increases For First Time In 100 Years." The Two-Way, National Public Radio. April 11, 2016.

p. 129 Against this backdrop, Wilson wrote

Wilson, E. O. The Creation: An Appeal to Save Life on Earth. New York: W. W. Norton & Company, 2006.

p. 130 The latest Gallup Poll shows that over 90 percent of US citizens believe in God

Newport, Frank. "More Than 9 in 10 Americans Continue to Believe in God." Gallup. June 3, 2011. http://www.gallup.com/poll/147887/americans-continue-believe-god.aspx.

Bohon, Dave. "Gallup: More Than 90 Percent of Americans Believe in God." New American, June 10, 2011.

p. 131 Sharks have been observed moving to deeper waters just before a hurricane strikes

Whitney, Nick. "Shark Behavior During Hurricanes: Not So 'Sharknado.'" Ocearch. August 26, 2013. http://www.ocearch.org/2013/08/26/shark-behavior-during-hurricanes-not-so-sharknado/.

"Nature's Forecasters." BBC Weather. April 23, 2008.

Boyle, Alan. "How Animals Gauge the Weather." NBC News Cosmic Log. February 1, 2008. http://cosmiclog.nbcnews.com/_news/2008/02/01/4349920-how-animals-gauge-the-weather.

Vatalaro, Michael. "Sharks' Sixth Sense." Boat U.S. Magazine, April 23, 2008.

p. 131 And there are many reports of animals escaping to safety before an earthquake and tsunami occur

"Can Animals Predict Disaster?" TV episode. Written and produced by Jeff Swimmer. Nature, Public Broadcasting System, 2008.

Sheldrake, Rupert. "Listen to the Animals: Why Did So Many Animals Escape December's Tsunami?" The Ecologist, March 2005.

Mott, Maryann. "Did Animals Sense Tsunami Was Coming?" National Geographic News. January 4, 2005. http://news.nationalgeographic. com/news/2005/01/0104_050104_tsunami_animals.html.

p. 131   Michelle Heupel, a scientist at the Mote Marine Laboratory

"Can Animals Predict Disaster? Tall Tales or True?" Nature, Public Broadcasting System. June 3, 2008. http://www.pbs.org/wnet/ nature/can-animals-predict-disaster-tall-tales-or-true/131/.

p. 132   In 2013, I had an opportunity to interview William Scott

Scott, William, interview by the author. The Costa Report. November 11, 2013.

Baird, Robert A. "Pyro-Terrorism—The Threat of Arson Induced Forest Fires as a Future Terrorist Weapon of Mass Destruction." Research paper, Marine Corps University School of Advanced Warfighting, 2005.

"'Forrest Fire Jihad' Could Pose a Significant Risk." SSI News. September 13, 2013. http://www.homelandsecurityssi.com/news/ entry/forrest-fire-jihad-could-pose-a-significant-risk.

Neale, Robert A. "Arson: The Overlooked Threat to Homeland Security." Emergency Management, September 7, 2010.

Ballam, Ed. "Feds: Pyro-Terrorism a Real Threat in the U.S." Officer. com. February 21, 2013. http://www.officer.com/news/10882958/ feds-pyro-terrorism-a-real-threat-in-the-us.

Gabbert, Bill. "USFS Deputy Director of Fire and Aviation Talks about Pyroterrorism." Wildfire Today. February 21, 2013. http:// wildfiretoday.com/2013/02/21/usfs-deputy-director-of-fire-and- aviation-talks-about-pyroterrorism/.

Gabbert, Bill. "Economic Warfare by Forest fire." Wildfire Today. September 8, 2012. http://wildfiretoday.com/2012/09/08/ economic-warfare-by-wildfires/.

p. 132   In 2011, the Navy Seal Six raided Osama Bin Laden's lair in Pakistan

Jolly, Dave. "Only You Can Prevent Terrorists from Starting Forest Fires." Washington Examiner, August 17, 2013.

Kovacs, Joe. "Are Terrorists Setting U.S. Wildfires?" WorldNetDaily.com. September 11, 2013. http://www.wnd.com/2013/09/are-terrorists-setting-u-s-wildfires/.

"'Unleash Hell': New Al Qaeda Magazine Describes in Detail How to Start Huge Forest Fires Across the U.S. with Instructions on How to Make 'Ember Bombs.'" DailyMail, May 3, 2012.

Tabankin, Ira. America on Fire. Amazon Digital Services, 2017.

Webster, Michael. "Are America's Enemies, the Terrorists, Behind the Yosemite and Other Western State Fires?" Examiner.com. September 3, 2013.

Gabbert. "Economic Warfare by forest fire."

"'Forrest Fire Jihad' Could Pose a Significant Risk."

p. 133 Then the director of Russia's Federal Security Bureau Aleksandr Bortnikov

Kovacs. "Are Terrorists Setting U.S. Wildfires?"

Tapscott, Mark. "Expert Warns Terrorists May Be Setting Wildfires Across American West." Washington Examiner, June 16, 2013.

CHAPTER SIX
The Invisible Tether

p. 139 A fire, a snake, the sound of footsteps behind us at night

"Understanding the Stress Response." Harvard Health Publications. Last updated March 18, 2016. http://www.health.harvard.edu/staying-healthy/understanding-the-stress-response.

Schoen, Marc. Your Survival Instinct Is Killing You: Retrain Your Brain to Conquer Fear and Build Resilience. New York: Plume, 2014.

Layton, Julia. "How Fear Works." How Stuff Works. September 13, 2005. http://science.howstuffworks.com/life/inside-the-mind/emotions/fear.htm.

p. 140   In 2016, news bureaus reported that our numbers would approach 7.5 billion within one year

World Population Clock. Worldometers. Accessed March 18, 2017. http://www.worldometers.info/world-population/.

Kunzig, Robert. "A World With 11 Billion People? New Projections Shatter Earlier Estimates." National Geographic, September 19, 2014.

Donaldson-Evans, Christine. "Experts: World Population Will Explode by 2025 with Influx of 'Megacities' of 10 Million People of More." Fox News. September 20, 2007. http://www.foxnews.com/story/2007/12/20/experts-world-population-will-explode-by-2025-with-influx-megacities-10-million.html.

US Census Bureau. "International Data Base." International Programs. Accessed January 11, 2012. https://web.archive.org/web/20120111060647/http:/www.census.gov/population/international/data/idb/region.php?N=%20Region%20Results%20

"Population Seven Billion: UN Sets Out Challenges." *BBC News*. October 26, 2011. http://www.bbc.com/news/world-15459643.

Coleman, Jasmine. "World's 'Seven Billionth Baby' Is Born." *The Guardian*. October 31, 2011. https://www.theguardian.com/world/2011/oct/31/seven-billionth-baby-born-philippines?intcmp=122.

"7 Billion People Is a 'Serious Challenge.'" *United Press International*. October 31, 2011. http://www.upi.com/Top_News/US/2011/10/31/7-billion-people-is-a-serious-challenge/UPI-73301320046200/.

"The World Population at 7 Billion." *U.S. Census Bureau* (blog). October 31, 2011. http://blogs.census.gov/2011/10/31/the-world-population-at-7-billion/.

p. 140   They noted it took 150,000 years to create the first billion humans. The second billion around 123 years

Radford, Tim. "How Many People Can the Earth Support?" *The Guardian,* November 11, 2004.

p. 141   The fact is, scientists estimate there is a greater than 80 percent probability we will hit between 9.6 to 12.3 Billion people by 2100

"World Population Projected to Reach 9.6 Billion by 2050—UN Report." *UN News Centre.* June 14, 2013. http://www.un.org/apps/news/story.asp?NewsID=45165#.WRNsLdLyvIU.

"World Population Prospects, the 2012 Revision: 'Low Variant' and 'High Variant' Values." United Nations. 2012. https://esa.un.org/unpd/wpp/.

"Proceedings of the United Nations Technical Working Group on Long-Range Population Projections." New York: United Nations Headquarters, June 30, 2003.

United Nations, Department of Economic and Social Affairs, Population Division. "World Population Prospects, 2015 Revision." Accessed May 10, 2017. https://esa.un.org/unpd/wpp/.

p. 141   Joel E. Cohen, the author of *How Many People Can the Earth Support?*

Cohen, Joel E. *How Many People Can the Earth Support?* New York: W. 141. Norton & Company, 1996.

Pearce, *Fred. The Coming Population Crash and Our Planet's Surprising Future.* Boston: Beacon Press, 2010.

Livi-Bacci, Massimo. *A Concise History of World Population.* West Sussex: Wiley-Blackwell, 2012.

Weisman, Alan. *Countdown: Our Last, Best Hope for a Future on Earth?* New York: Back Bay Books, 2014.

Ehrlich, Paul R., and Anne H. Ehrlich. *Population, Resources, Environment: Issues in Human Ecology.* London: W.H. Freeman, 1970.

Daily, Gretchen C, Paul R. Ehrlich, and Anne H. Ehrlich. "Optimum Human Population Size." *Population and Environment: A Journal of Interdisciplinary Studies* 15, no. 6(1994).

p. 141   The total combined land mass of Earth only adds up to roughly 15 Billion hectares

Cohen. *How Many People Can the Earth Support?*

Andregg, Michael M. *Seven Billion and Counting: The Crisis in Global Population Growth.* Minneapolis: Twenty-First Century Books, 2014.

Emmott, Stephen. *Ten Billion.* New York: Vintage Books, 2013.

Weisman. *Countdown: Our Last, Best Hope for a Future on Earth?*

Ehrlich and Ehrlich. *Population, Resources, Environment: Issues in Human Ecology.*

Daily, Ehrlich, and Ehrlich. "Optimum Human Population Size."

p. 142   Thanks to these and other innovations, agricultural productivity in the United States is three to four times greater than it was in 1948

Wang, Sun Ling, Paul Heisey, David Schimmelpfennig, and Eldon Ball. *Agricultural Productivity Growth in the United States: Measurement, Trends and Drivers.* Economic Research Report No. (ERR-189), USDA, July 2015.

p. 142   Global yields for rice—a main food staple—have not improved since 1966

Ray, Deepak K., Nathaniel D. Mueller, Paul C. West, and Jonathan A. Foley. "Yield Trends Are Insufficient to Double Global Crop Production by 2050." *PLoS ONE* 8, no. 6(2013). doi: 10.1371/journal.pone.0066428.

Ricepedia.org

Brown, Lester R. *Full Planet, Empty Plates: The New Geopolitics of Food Scarcity*. New York: W. W. Norton & Company, 2012.

p. 142   In 1950, broccoli had 130 mg of calcium. Today, it has less than 48 mg

Summer. "Fruits and Vegetables Less Healthy Now Than 50 Years Ago." *Growing Your Baby*, July 14, 2010. http://www. growingyourbaby.com/2010/07/14/fruits-and-vegetables-less-healthy-now-than-50-years-ago/

Ray, Claiborne C. "A Decline in the Nutritional Value of Crops." *New York Times*, September 12, 2015.

p. 143   Today, more than five million people sleep covered with no more than a newspaper on the streets of Mumbai, India

Tatke, Sukhada. "Homeless in a Wet City." *The Hindu*, August 5, 2013.

UN-Habitat. *The Challenge of Slums: Global Report on Human Settlements 2003*. New York: UN-Habitat, 2003.

UN-Habitat. "State of the World's Cities 2008/2009." New York: UN-Habitat, 2008.

Tipple, Graham, and Suzanne Speak. "Definitions of Homelessness in Developing Countries." *Habitat International* 29, no. 2 (2005): 337–352. doi: 10.1016/j.habitatint.2003.11.002.

p. 143   The latest estimates put global homelessness at two hundred million people

United Nations, Department of Economic and Social Affairs, Statistics Division, Demographic and Social Statistics Branch. "United Nations *Demographic Yearbook* Review: National Reporting of Household Characteristics, Living Arrangements and Homeless Households: Implications for International Recommendations." New York: United Nations, 2004.

"Enumeration of Homeless People." Presentation at the Twelfth Meeting of the United Nations Economic and Social Council,

Economic Commission for Europe Conference of European Statisticians, Group of Experts on Population and Housing Censuses, Geneva, October 28–30, 2009. https://unstats.un.org/unsd/censuskb20/Attachments/2009MPHASIS_ECE_Homeless-GUID25ae612721cc4c2c87b536892e1ed1e1.pdf.

Zarocostas, John. "Homelessness Increasing All Over the World." *Washington Times*, April 11, 2005. http://www.washingtontimes.com/news/2005/apr/10/20050410-105739-5991r/.

Zimmerman, Kayla. "A Global Crisis: Homelessness." *Prezi*. May 5, 2014. https://prezi.com/11drljpp9bpk/a-global-crisis-homelessness/.

Capdevila, Gustavo. "Human Rights: More Than 100 Million Homeless Worldwide." *Inter Press Service*. March 30, 2005. http://www.ipsnews.net/2005/03/human-rights-more-than-100-million-homeless-worldwide/.

YXC Project, UNEP/UNESCO. "Homeless: Developing Countries." *Youth Xchange*. Accessed May 10, 2017. http://www.youthxchange.net/main/b236_homeless-i.asp.

p. 143   Last year the United Nations reported the number of displaced persons hit a new high

McKirdy, Euan. "UNHCR Report: More Displaced Now Than After WWII." CNN. June 20, 2016.

Internal Displacement Monitoring Center. "Grid 2016: Global Report on Internal Displacement. *IDMC*. Accessed May 10, 2017. http://www.internal-displacement.org/globalreport2016/.

"Number Displaced Worldwide Hits Record High – UN Report." BBC. June 18, 2015.

p. 143   Since 1855, when the first rubber condom was produced

"History of Contraception: Condoms and Sponges." Case Western Reserve University. Accessed May 10, 2017. http://case.edu/affil/skuyhistcontraception/online-2012/Condoms-Sponges.html.

DeNoon, Daniel. "Birth Control Timeline." *Medicine.net.* July 17. 2003. http://www.medicinenet.com/script/main/art.asp?articlekey=52188.

Alef, Daniel. *Charles Goodyear: The Great Vulcanizer.* Audiobook. Santa Barbara: Meta4 Press, 2009.

Lord, Alexandra M. *Condom Nation: The U.S. Government's Sex Education Campaign from World War I to the Internet.* Baltimore: Johns Hopkins University Press, 2010.

p. 147 At the end of 2010, the suicide of a twenty-six-year-old Tunisian street vendor ignited the biggest revolution of the twenty-first century

Worth, Robert F. "How a Single Match Can Ignite a Revolution." *New York Times.* January 21, 2011. http://www.nytimes.com/2011/01/23/weekinreview/23worth.html.

Beaumont, Peter. "Mohammed Bouazizi: The Dutiful Son Whose Death Changed Tunisia's Fate." *The Guardian,* January 20, 2011.

"Turmoil in Tunisia: As It Happened on Friday." BBC News. January 14, 2011.

von Rohr, Mathieu. "The Small Tunisian Town that Sparked the Arab Revolution." *Spiegel Online.* March 18, 2011. http://www.spiegel.de/international/world/the-fruits-of-mohamed-the-small-tunisian-town-that-sparked-the-arab-revolution-a-751278.html.

Bodden, Valerie. *The Arab Spring.* Mankato, Minn.: Creative Paperbacks, 2017.

Ryan, Yasmine. "The Tragic Life of a Street Vendor." *Al Jazeera.* January 20, 2011. http://www.aljazeera.com/indepth/features/2011/01/201111684242518839.html.

"Tunisian Protester Dies of Burns." *Al Jazeera.* January 5, 2011. http://www.aljazeera.com/news/africa/2011/01/201115101926215588.html.

Fahim, Kareem. "Slap to a Man's Pride Set Off Tumult in Tunisia." *New York Times*, January 21, 2011. http://www.nytimes.com/2011/01/22/world/africa/22sidi.html?_r=1&p. wanted=2&src=twrhp.

Watson, Ivan, and Jomana Karadsheh. "The Tunisian Fruit Seller Who Kickstarted Arab Uprising." CNN. March 22, 2011. http://www.cnn.com/2011/WORLD/meast/03/22/tunisia.bouazizi.arab.unrest/.

Abouzeid, Rania. "Bouazizi: The Man Who Set Himself and Tunisia on Fire." *Time*, January 21, 2011. http://content.time.com/time/magazine/article/0,9171,2044723,00.html.

Li, Hao. "The Story of Mohamed Bouazizi, the Man Who Toppled Tunisia." *International Business Times*. January 14, 2011. http://www.ibtimes.com/story-mohamed-bouazizi-man-who-toppled-tunisia-255077.

Haas, Mark L., and David W. Lesch. *The Arab Spring: The Hope and Reality of the Uprisings*. Boulder, Colo.: Westview Press, 2017.

Ghanem, Hafez. *The Arab Spring Five Years Later: Toward Great Inclusiveness*. Washington, D.C.: The Brookings Institution, 2016.

p. 150   By January 2011 the revolution had succeeded

Chrisafis, Angelique, and Ian Black. "Zine al-Abidine Ben Ali Forced to Flee Tunisia as Protesters Claim Victory." *The Guardian*, January 14, 2011.

Associated Press. "Morocco Announces Constitutional Reform Plan." *The Guardian*, March 9, 2011.

Mihalakas, Nasos. "Constitutional Reforms in Morocco and Jordan!" *Foreign Policy Blogs*. September 1, 2011. http://foreignpolicyblogs.com/2011/09/01/constitutional-reforms-in-morocco-and-jordan/.

"Middle East Unrest: Saudi and Bahraini Kings Offer Concessions." *The Guardian*, February 23, 2011. https://www.theguardian.com/world/2011/feb/23/middle-east-unrest-concessions.

"Moroccan Monarch Pledges Reform." *Al Jazeera*. March 9, 2011. http://www.aljazeera.com/news/africa/2011/03/201139204839521962.html.

Karam, Zeina. "Syria to Consider Constitutional Reform, Says President." *Toronto Star*, June 20, 2011.

"Egypt Crisis: President Hosni Mubarak Resigns as Leader." *BBC News*. February 12, 2011.

Finn, Peter. "The Rise and Fall of Libyan Leader Moammar Gaddafi." *Washington Post*, August 25, 2011.

Bakri, Nada. "Bahrain Opposition Says King's Measures Fall Short." *New York Times,* January 15, 2012.

p. 151   Using Egypt as an example, in 1981

Worldpopulaitonhistory.org

Worldometers.com

p. 153   Estimates vary on how much money

Tomlinson, Simon. "Revealed: How Immigrants in America Are Sending $120 Billion to Their Struggling Families Back Home." *Daily Mail*, January 31, 2013.

p. 155   In 2010, the US Department of Transportation reported that cell phones were involved in 1.6 million accidents and more than six thousand deaths

Hanes, Stephanie. "Report: Cell Phone Distraction Causes One in Four U.S. Car Crashes." *Christian Science Monitor*, January 12, 2010.

Speaker, Burke. "Texting and Driving Statistics 2014." *InvestorPlace*. June 12, 2014. http://investorplace.com/2014/06/texting-driving-statistics-2014/#.WROJANLyvIU.

p. 156   In 2011, it took the fifth air traffic controller falling asleep on the job to prompt the Federal Aviation Administration to admit they had a problem with fatigue

Stark, Lisa, and Kevin Dolak. "Fifth Incident of Sleeping Air Traffic Controller Prompts Schedule Actions." ABC News. April 16, 2011.

Alpert, Lucas. "FAA Air Traffic Control Chief Hank Krakowski Resigns After ANOTHER Controller Falls Asleep in Tower." *Daily News*, April 14, 2011.

Austin, Anastacia Mott. "Americans Just Not Getting Enough Sleep, Study Shows."

Buzzle.*com*. May 7, 2009. www.buzzle.com.

Lowry, Joan. "Air Traffic Controllers Working Exhausting Schedules, Report Finds." PBS News Hour. June 13, 2014.

p. 156   When asked about the number of hours

Stark, Lisa, and Kevin Dolak. "Fifth Incident of Sleeping Air Traffic Controller Prompts Schedule Actions." ABC News. April 16, 2011.

p. 157   According to the Pew Research Center, 27 percent of adults between the ages of eighteen and twenty-four use online dating

Smith, Aaron, and Monica Anderson. "5 Facts About Online Dating." Pew Research Center. February 29, 2016. http://www.pewresearch. org/fact-tank/2016/02/29/5-facts-about-online-dating/.

p. 157   The company quickly became the best-backed start-up in the world, raising nearly $1 billion in cash at their initial public offering

Richtel, Matt. "Webvan Stock Price Closes 65% Above Initial Offering." *New York Times*, November 6, 1999. http://www. nytimes.com/1999/11/06/business/webvan-stock-price-closes-65- above-initial-offering.html.

Anders, G. "How Webvan Conquers E-Commerce's Last Mile." *Wall Street Journal*, December 15, 1999.

Goldman, David. "10 Big Dot-Com Flops." *CNNMoney*. March 2, 2015. http://money.cnn.com/gallery/technology/2015/03/02/dot- com-flops/2.html.

Lanxon, Nate. "The Greatest Defunct Web Sites and Dotcom Disasters." *CNET*. November 18, 2009. https://www.cnet.com/ uk/news/the-greatest-defunct-web-sites-and-dotcom-disasters/.

"Webvan Bags It for Good." *ZDNet*. July 9, 2001. http://www.zdnet. com/article/webvan-bags-it-for-good-5000117745/.

Sandoval, Greg. "Webvan Delivers Its Last Word: Bankruptcy." *CNET.* September 7, 2007. https://www.cnet.com/news/webvan-delivers-its-last-word-bankruptcy/.

Wolverton, Troy. "Seeking Relics Amid Webvan's Ruins." *CNET.* October 30, 2001. https://www.cnet.com/news/seeking-relics-amid-webvans-ruins/.

Beltran, Luisa. "Webvan Goes Shopping." *CNNMoney.* June 26, 2000. http://money.cnn.com/2000/06/26/deals/webvan/.

Emert, Carol. "Venture Lessons in Webvan Collapse/Financing History a Cautionary Tale." *San Francisco Chronicle*, July 15, 2001. http://www.sfgate.com/bayarea/article/Venture-lessons-in-Webvan-collapse-Financing-2899418.php.

Delgado, Ray. "Webvan Goes Under/Online Grocer Shuts Down—$830 Million Lost, 2,000 Workers Fired." *San Francisco Chronicle*, July 9, 2001. http://www.sfgate.com/news/article/Webvan-goes-under-Online-grocer-shuts-down-2901586.php.

p. 158   Then, five years after opening their doors, and eighteen months after their IPO

Burgess, Kristy, Sharon T Hopkins, and Kenneth White."Webvan: A Cautionary Tale." Dissertation, Piedmont University, 2006.

Goldman. "10 Big Dot-Com Flops."

Lanxon. "The Greatest Ddefunct Web Sites and Dotcom Disasters."

Wolverton. "Seeking Relics Amid Webvan's Ruins."

Emert. "Venture Lessons in Webvan Collapse/Financing History a Cautionary Tale."

"Webvan Bags It for Good."

Sandoval. "Webvan Delivers Its Last Word: Bankruptcy."

Blackwell, R. "Why Webvan Went Bust." *Wall Street Journal*, July 16, 2001.

Evans, B. "Webvan: Who's to Blame?" *Information Week*, July 2001.

Helft, M. "The End of the Road." *The Industry Standard*, July 23, 2001.

Weber, J. "The Fall of Webvan." *The Industry Standard*, July 23, 2001.

CHAPTER SEVEN
Bioleverage

p. 165   And he'd have to put his faith in two engineers from the Ryan Aeronautical Company—fellows that outfitted the plane in less than sixty days

Bak, Richard. *The Big Jump: Lindbergh and the Great Atlantic Air Race.* Hoboken, NJ: John Wiley & Sons, 2011.

Pryor, Alton. *Charles Lindbergh: The Rogue Aviator.* Roseville, CA: Stagecoach Publishing, 2013.

Hampton, Dan. *The Flight: Charles Lindbergh's Daring and Immortal 1927 Transatlantic Crossing.* New York: William Morrow, 2017.

Van Der Linden, F. Robert, Dominick A. Pisano, and Reeve Lindbergh. *Charles Lindbergh and the Spirit of St. Louis.* New York: Harry N. Abrams, 2002.

p. 165   But he owed the State National Bank $15,000

Bak. *The Big Jump: Lindbergh and the Great Atlantic Air Race.*

Pryor. *Charles Lindbergh: The Rogue Aviator.*

Hampton. *The Flight: Charles Lindbergh's Daring and Immortal 1927 Transatlantic Crossing.*

Van Der Linden, Pisano, and Lindbergh. *Charles Lindbergh and the Spirit of St. Louis.*

p. 166   Thus began Charles Lindbergh's 1927 solo flight from New York to Paris. It took the pilot 30 hours to cross 3,600 miles

Bak. *The Big Jump: Lindbergh and the Great Atlantic Air Race.*

Pryor. *Charles Lindbergh: The Rogue Aviator.*

Hampton. *The Flight: Charles Lindbergh's Daring and Immortal 1927 Transatlantic Crossing.*

Van Der Linden, Pisano, and Lindbergh. *Charles Lindbergh and the Spirit of St. Louis.*

p. 166   So by the time The Spirit of St. Louis touched ground at Le Bourget Airport

Hampton. *The Flight: Charles Lindbergh's Daring and Immortal 1927 Transatlantic Crossing.*

Van Der Linden, Pisano, and Lindbergh. *Charles Lindbergh and the Spirit of St. Louis.*

Bak. *The Big Jump: Lindbergh and the Great Atlantic Air Race.*

Pryor. *Charles Lindbergh: The Rogue Aviator.*

p. 166   But in the days following Lindbergh's historic achievement, the man behind modern air travel, financier and hotelier Raymond Orteig, was forgotten

Sparks, Evan. "Raymond Orteig." The Philanthropy Hall of Fame, *Philanthropy Roundtable.* Accessed May 10, 2017. http://www.philanthropyroundtable.org/almanac/hall_of_fame/raymond_orteig.

Bak. *The Big Jump: Lindbergh and the Great Atlantic Air Race.*

Diamandis, Peter, and Steven Kotler. *Abundance: The Future Is Better Than You Think.* New York: Free Press, 2012.

Jackson, Joe. *Atlantic Fever: Lindbergh, His Competitors, and the Race to Cross the Atlantic.* New York: Picador, 2012.

Pryor. *Charles Lindbergh: The Rogue Aviator.*

Hampton. *The Flight: Charles Lindbergh's Daring and Immortal 1927 Transatlantic Crossing.*

Van Der Linden, Pisano, and Lindbergh. *Charles Lindbergh and the Spirit of St. Louis.*

p. 168   After all, it was the French Academy's ₣100,000 (francs) reward

Aftalion, Fred. *A History of the International Chemical Industry*. Philadelphia: University of Pennsylvania Press, 1991.

Tietz, Tabea. "Nicholas Leblanc and the Leblanc Process." *SciHi Blog*. December 2014.

"Nicolas Leblanc: French Chemist." *Encyclopedia Britannica*. July 20, 1998. https://www.britannica.com/biography/Nicolas-Leblanc.

p. 168   And the British Parliament's Longitude Prize was the impetus for John Harrison to perfect the chronometer

longitudeprize.org

Roberts, Alice. "A True Sea Shanty: The Story Behind the Longitude Prize." *The Guardian*, May 17, 2014.

Carter, William E., and Merri Sue Carter. "The British Longitude Act Reconsidered." *American Scientist*, April 19, 2015. http://www.americanscientist.org/issues/pub/the-british-longitude-act-reconsidered.

Gould, Rupert T. *The Marine Chronometer. Its History and Development*. London; J. D. Potter, 1923.

"John Harrison: Timekeeper to Nostell and the World!" *BBC*. April 8, 2009. http://www.bbc.co.uk/bradford/content/articles/2009/04/06/nostell_john_harrison_feature.shtml.

p. 168   In each case, up to one hundred times the actual prize money was invested in the pursuit of victory.

Diamandis, Peter. "The ROI of Prize Incentives." *Stanford.edu eCorner*. April 23, 2008. http://ecorner.stanford.edu/videos/2004/The-ROI-of-Prize-Incentives.

Diamandis and Kotler. *Abundance: The Future Is Better Than You Think*.

Diamandis, Peter H., and Steven Kotler. *Bold: How to Go Big, Make Bank, and Better the World*. New York: Simon & Schuster Audio, 2015. Audio CD.

Dean, Jason. "Peter Diamandis: Eyes on the XPRIZE." *CSQ*, January 1, 2015.

Xprize.org

p. 169   Armed with these insights, in 1996—on the forty-third anniversary of Alan Shepard's first spaceflight

Leonard, David. "SpaceShipOne Wins $10 Million Ansari X Prize in Historic 2nd Trip to Space." *Space.com.* October 4, 2004. http://www.space.com/403-spaceshipone-wins-10-million-ansari-prize-historic-2nd-trip-space.html.

Belfiore, Michael. *Rocketeers: How a Visionary Band of Business Leaders, Engineers, and Pilots Is Boldly Privatizing Space.* New York: Smithsonian Books, 2007.

Diamandis and Kotler. *Bold: How to Go Big, Make Bank, and Better the World.*

"Ansari," *XPrize,* http://ansari.xprize.org/

p. 169   Then, in 2004, Burt Rutan laid claim to the $10 million prize

Leonard. "SpaceShipOne Wins $10 Million Ansari X Prize in Historic 2nd Trip to Space."

Boyle, Alan. "Spaceship Team Gets Its $10 Million Prize." *NBC News.* June 16, 2011. http://www.nbcnews.com/id/6421889/ns/technology_and_science-space/t/spaceship-team-gets-its-million-prize/#.WROXsdLyvIU.

"SpaceShipOne Rockets to Success." *BBC News.* October 28, 2010. http://news.bbc.co.uk/2/hi/science/nature/3712998.stm.

Jefferson, Catherine A. "First Private Manned Space Flight." *Devsite. org.* October 8, 2007. https://web.archive.org/web/20071008125717/http:/www.devsite.org/lab/spaceshipone/data.shtml.

Guthrie, Julian. *How to Make a Spaceship: A Band of Renegades, an Epic Race, and the Birth of Private Spaceflight.* New York: Penguin Press, 2016.

p. 169   At that time, NASA's Space Shuttle program cost taxpayers $200 billion

"SpaceShipOne Rockets to Success."

Howell, Elizabeth. "Paul Allen: Billionaire Backer of Private Space Ventures." *Space.com.* January 18, 2013. http://www.space.com/19333-paul-allen.html.

Foust, Jeff. "Paul Allen's Past (and Future) in Space." *The Space Review,* April 25, 2011.

Belfiore. *Rocketeers: How a Visionary Band of Business Leaders, Engineers, and Pilots Is Boldly Privatizing Space.*

p. 170   Virgin Galactic tickets went on sale for $200,000 a passenger for the once-in-a-lifetime opportunity to travel seventy miles above the earth's surface

Chang, Alicia. "Dream Is Over for Virgin Galactic Space Tourist." *Yahoo! News.* October 4, 2011. https://www.yahoo.com/news/dream-over-virgin-galactic-space-tourist-073153281.html.

"Virgin Sells Space Flight Tickets for $200,000." *Reuters,* July 18, 2006.

p. 171   Oil-skimming technology was so inadequate and outdated

Nalder, Eric. "Decades After Exxon Valdez, Cleanup Technology Still Same." *Houston Chronicle,* May 17, 2010.

Fountain, Henry. "Advances in Oil Spill Cleanup Lag Since Valdez." *New York Times,* June 24, 2010.

p. 171   So Schmidt joined forces with Diamandis

Boyle, Rebecca, and Clay Dillow. "Winner of Million-Dollar X Challenge Cleans Up Oil Spills Three Times Better Than Existing Tech." *Popular Science,* October 11, 2011. http://www.popsci.com/technology/article/2011-10/x-prize-challenge-seeks-new-designs-and-inspiration-clean-future-oil-spills.

Harrington, Kent. "The $1 Million Wendy Schmidt Oil Cleanup X-Prize Winners Announced." *ChEnected.* December 16, 2011. https://www.aiche.org/chenected/2011/12/1-million-wendy-schmidt-oil-cleanup-x-prize-winners-announced.

p. 172   In October 2011, the Elastec/American Marine Company of Illinois won Schmidt's X Prize

Boyle and Dillow. "Winner of Million-Dollar X Challenge Cleans Up Oil Spills Three Times Better Than Existing Tech."

Harrington. "The $1 Million Wendy Schmidt Oil Cleanup X-Prize Winners Announced."

Woody, Todd. "Wendy Schmidt's X Prize Oil Cleanup Challenge Names Winners." *Forbes*, October 11, 2011.

p. 172   Stewart and Marilyn Blusson, the sponsors of the Archon Genomics X Prize

"Archon Genomics," *XPrize*, http://genomics.xprize.org/

p. 174   Salinas' homicide rate was running five times the national average

Anderson, Marla O., Honorable Judge. "City of Salinas' Responses to Monterey County Civil Grand Jury Final Report No. 10 and Final Report no. 11." Salinas, CA, August 19, 2014.

"Salinas: A deadly place to live." *NBC News,* August 21, 2008.

Reynolds, Julia. "What's Behind the Salinas Homicide Rate?" *Monterey Herald/Mercury News*, December 28, 2015.

p. 178   With financial help from the federal government, New York City undertook one of the largest

Von Essen, Tom. "How New York Became Safe: The Full Story." *City Journal*, Special Issue 2009.

Associated Press. "Did Giuliani Really Clean Up Times Square?" *Free Republic* via CBS News. December 28, 2007. http://www.freerepublic.com/focus/f-news/1945378/posts.

Bratton, William, and Peter Knobler. *Turnaround: How America's Top Cop Reversed the Crime Epidemic.* New York: Random House, 1998.

Dussault, Raymond. "Jack Maple: Betting on Intelligence." *Government Technology*, March 31, 1999.

Langan, Patrick A. and Matthew R. Durose, "The Remarkable Drop in Crime in New York City." *Bureau of Justice Statistics*. October 21, 2004.

Greene, Judith. "Zero Tolerance: A Case Study of Police Policies and Practices in New York City." *Crime and Delinquency* 45, no. 2 (1999): 171–187. doi: 10.1177/0011128799045002001.

Barrett, Wayne. *Rudy! An Investigative Biography of Rudolph Giuliani*. New York: Basic Books, 2000.

MacDonald, Heather. "New York Cops: Still the Finest." *City Journal*, Summer 2006. https://www.city-journal.org/html/new-york-cops-still-finest-12951.html.

Zimring, Franklin E. *The Great American Crime Decline*. Oxford: Oxford University Press, 2007.

Lacayo, Richard. "Crime: Law and Order." *Time*, January 15, 1996. http://content.time.com/time/magazine/article/0,9171,983958,00.html.

p. 179   While crime across the United States decreased by 30 percent

Von Essen. "How New York Became Safe: The Full Story."

Associated Press. "Did Giuliani Really Clean Up Times Square?"

Bratton and Knobler. *Turnaround: How America's Top Cop Reversed the Crime Epidemic.*

Dussault. "Jack Maple: Betting on Intelligence."

Langan and Durose. "The Remarkable Drop in Crime in New York City."

Greene. "Zero Tolerance: A Case Study of Police Policies and Practices in New York City."

Zimring. *The Great American Crime Decline.*

Lacayo. "Crime: Law and Order."

p. 182   Then slowly Salinas turned a corner

"Salinas, California." *City-Data.com*. Accessed May 10, 2017.
http://www.city-data.com/city/Salinas-California.html.

p. 183   In 1964, W.D. Hamilton put forth such a theory

Hamilton, W. D. "The Evolution of Altruistic Behavior." *American
Naturalist* 97 (1963): 354–356.

Hamilton, W. D. "The Genetical Evolution of Social Behaviour."
Journal of Theoretical Biology 7 (1964): 1–16.

**p. 183   Researchers were finding species of insects that shared a
lot of genetic material**

Luskin, Casey. "E.O. Wilson Disavows His Own Kin Selection."
Evolution News. May 4, 2011. https://www.evolutionnews.org/2011/
05/eo_wilson_disavows_his_own_kin/.

Nowak, Martin, Corina Tarnita, and Edward O. Wilson. "The
Evolution of Eusociality." *Nature* 466, no. 7310 (2010): 1057–1062.
doi: 10.1038/nature09205.

Wilson, Edward O. *The Social Conquest of Earth*. New York: Liveright
Publishing Corporation, 2013.

Queller, David C., and Joan E. Strassman. "Quick Guide:
Kin Selection." *Current Biology* 12, no. 24 (2002): R832. doi:
10.1016/S0960-9822(02)01344-1.

Abbot, Patrick, et al. "Inclusive Fitness Theory and Eusociality."
Nature 471, no. 7339 (2011): 10. doi: 10.1038/nature09831.

p. 184   In his Pulitzer-prize winning book *On Human Nature*

Wilson, Edward O. *On Human Nature*. Cambridge, MA: Harvard
University Press, 1978.

p. 185   Author Robert Wright observes

Wright, Robert. *The Moral Animal: Why We Are the Way We Are:
The New Science of Evolutionary Psychology*. New York: Vintage
Books, 1995.

CHAPTER EIGHT
A Prescient Mind

p. 194   Dr. Wolfram Schultz, neuroscientist at the University of Cambridge, is largely responsible for the discovery of prediction neurons

Schultz, Wolfram, and Anthony Dickinson. "Neuronal Coding of Prediction Errors." *Annual Review of Neuroscience* 23(2000): 473–500. doi: 10.1146/annurev.neuro.23.1.473.

Fiorillo, Christopher D., William T. Newsome, and Wolfram Schultz. "The Temporal Precision of Reward Prediction in Dopamine Neurons." *Nature Neuroscience* 11 (2008): 966–973. doi:10.1038/nn.2159.

Bayer, Hannah M., and Paul W. Gilmcher. "Midbrain Dopamine Neurons Encode Quantitative Reward Prediction Error Signal." *Neuron* 47, no. 1 (2005): 129–141. doi: 10.1016/j.neuron.2005.05.020.

Shidara, Munetaka, and Barry J. Richmond. "Anterior Cingulate: Single Neuronal Signals Related to Degree of Reward Expectancy." *Science* 296, no. 5573 (2002): 1709–1711. doi: 10.1126/science.1069504.

Marzullo, Timothy C., Edward G. Rantz, and Gregory J. Gage. "Stock Market Behavior Predicted by Rat Neurons." *Annals of Improbable Research* 12, no. 4 (2006): 22–25.

p. 196   So impressive that Jeffrey Zacks, researcher at Washington University

"Everyday Clairvoyance: How Your Brain Makes Near-Future Predictions." *EurekAlert.* August 17, 2011. https://www.eurekalert.org/pub_releases/2011-08/wuis-echo81711.php.

Smith, Graham. "We CAN Predict the Future (a Bit): Why the Brain Knows What's Going to Happen Before It Does." *Daily Mail*, August 23, 2011. http://www.dailymail.co.uk/sciencetech/

article-2029189/We-CAN-predict-future-The-brain-knows-whats-going-happen-does.html#ixzz1VreOQzkI.

Zacks, Jeffrey M., et al. "Event Perception: A Mind-Brain Perspective." *Psychological Bulletin* 133, no. 2 (2007): 273–293.

Taylor, Kimberly Hayes. "Forget Psychic Hotline. You Can Predict the Near-Future Everyday." *The Body Odd*, NBC News. August 22, 2011. http://bodyodd.nbcnews.com/_news/2011/08/22/7420028-forget-psychic-hotline-you-can-predict-the-near-future-everyday.

p. 197   Meanwhile, at the University of California at Los Angeles

Wolpert, Stuart. "Neuroscientists Can Predict Your Behavior Better Than You Can." *Science Daily*, June 22, 2010.

p. 197   Kent Kiehl, a neuroscientist, at the Mind Research Network

Aharonia, Eyal, et al. "Neuroprediction of Future Rearrest." *Proceedings of National Academy of Sciences* 110, no. 15 (2013): 6223–6228. doi: 10.1073/pnas.1219302110.

Nuzzo, Regina. "Brain Scans Predict Which Criminals Are More Likely to Reoffend." *Nature News*, March 25, 2013. http://www.nature.com/news/brain-scans-predict-which-criminals-are-more-likely-to-reoffend-1.12672.

Haederle, Michael. "Brain Function Tied to Risk of Criminal Acts." *Los Angeles Times*, July 15, 2013.

p. 198   The area of the brain responsible for impulse control

Aharonia et al. "Neuroprediction of Future Rearrest."

Castillo, Michelle. "Brain Scans May Show Which Criminals Are More Likely to Continue Life of Crime." CBS *News*. March 27, 2013. http://www.cbsnews.com/news/brain-scans-may-show-which-criminals-are-more-likely-to-continue-life-of-crime/.

p. 198   According to Essi Viding

Castillo. "Brain Scans May Show Which Criminals Are More Likely to Continue Life of Crime."

p. 198   And clinical psychologist Dustin Pardini added, "It's a great study because it brings neuro-imaging into the realm of prediction"

Castillo. "Brain Scans May Show Which Criminals Are More Likely to Continue Life of Crime."

Shidara and Richmond. "Anterior Cingulate: Single Neuronal Signals Related to Degree of Reward Expectancy."

p. 199   And just as they anticipated—with no clues to go off except for the strength of P300 readings

Tremmel, Pat Vaughan. "Reading P300 Brain Waves to Predict Terrorist Attacks." *Neuroscience News*. July 31, 2010. http://neurosciencenews.com/reading-terrorists-p300-brain-waves-attacks/.

Meixner, John B., and J. Peter Rosenfeld. "A Mock Terrorism Application of the P300-Based Concealed Information Test." *Psychophysiology* 48, no 2 (2011): 149–154. doi: 10.1111/j.1469-8986.2010.01050.x.

Harrell, Eben. "Fighting Crime by Reading Minds." *Time*, August 7, 2010.

Ackerman, Spencer. "How to Catch a Terrorist: Read His Brainwaves—Really?" *Wired*, September, 10, 2010.

p. 200   Doctors Eleanor Maguire and Katherine Woollett from University College London set out to find out

Maguire, Eleanor A., Katherine Woollett, and Hugo J. Spiers. "London Taxi Drivers and Bus Drivers: A Structural MRI and Neuropsychological Analysis." *Hippocampus* 16 (2006): 1091–1101.

Jabr, Ferris. "Cache Cab: Taxi Drivers' Brains Grow to Navigate London's Streets."

Scientific American, December 8, 2011.

Brown, Mark. "How Driving a Taxi Change London Cabbie's Brains." *Wired*, December 9, 2011.

p. 201 Due to the degree of difficulty involved, in 2000, Maguire and Woollett began examining the brains of seventy-nine driver trainees

Maguire, Woollett, and Spiers. "London Taxi Drivers and Bus Drivers: A Structural MRI and Neuropsychological Analysis."

Jabr. "Cache Cab: Taxi Drivers' Brains Grow to Navigate London's Streets."

Brown. "How Driving a Taxi Change London Cabbie's Brains."

p. 202 Or as Dr. Maguire put it

"Taxi Drivers' Brains 'Grow' on the Job." *BBC News.* March 14, 2000. http://news.bbc.co.uk/2/hi/677048.stm.

p. 202 In April 2013, the President of the United States announced a government-sponsored program to "map the human brain"

Markoff, John. "Obama Seeking to Boost Study of Human Brain." *New York Times*, February 17, 2013. http://www.nytimes.com/2013/02/18/science/project-seeks-to-build-map-of-human-brain.html?p. wanted=all&_r=0.

Szalavitz, Maia. "Brain Map: President Obama Proposes First Detailed Guide of Human Brain Function." *Time*, February 19, 2013. http://healthland.time.com/2013/02/19/brain-map-president-obama-proposes-first-detailed-guide-of-human-brain-function/.

Kalil, Tom. "A White House Call To Action to Advance the BRAIN Initiative." U.S. White House Briefing. February 24, 2014. https://obamawhitehouse.archives.gov/blog/2014/02/24/white-house-call-action-advance-brain-initiative.

Alivisatos, A. Paul, et al. "The Brain Activity Map Project and the Challenge of Functional Connectomics." *Neuron* 74, no. 6 (2012): 970–974.

Markoff, John; Gorman, James. "Obama to unveil initiative to map the human brain". *New York Times*. April 2, 2013.

Fox, Maggie. "White House Pitches Brain Mapping Project." *NBC News*. April 2, 2013. http://vitals.nbcnews.com/_news/2013/04/02/17565983-white-house-pitches-brain-mapping-project.

"White House Neuroscience Initiative." *Whitehouse.gov*. May 2015. https://www.whitehouse.gov/administration/eop/ostp/initiatives

National Institutes of Health. "NIH Blueprint for Neuroscience Research." *NIH.gov*. May 7, 2015. https://neuroscienceblueprint.nih.gov/.

Mingming, Du, and Gao Yinan. "China Brain Project to be Launched." *People's Daily Online*. June 30, 2015. http://en.people.cn/n/2015/0630/c98649-8913112.html.

"Japanese Research Organizations Contribute to Human Brain Project." Okinawa Institute of Science and Technology. Press Release. January 29, 2013

"Announcement of the US-JAPAN Brain Research Cooperative Program—US Component." Notice Number: NOT-NS-07-009. May 3, 2007.

CHAPTER NINE
ForeWorld

p. 205   On January 6, 2015, humankind passed an important milestone. Astronomers catalogued the one-thousandth planet in space

Wall, Mike. "*NASA's Kepler Telescope Finds 1000th Alien Planet.*" *Christian Science Monitor, January 8, 2015.*

Phillips, Tony. "Kepler Discovers 1000th Exoplanet." *NASA Science Beta.* January 6, 2015. https://science.nasa.gov/science-news/science-at-nasa/2015/06jan_kepler1000.

NASA, and World Spaceflight News. *Complete Guide to the Kepler Space Telescope Mission and the Search for Habitable Planets and Earth-like Exoplanets: Planet Detection Strategies, Mission History and Accomplishments.* Progressive Management, 2013.

p. 205   That same day, scientists at NASA announced the Kepler spacecraft had "compiled a list of 3,500 more candidates." Thirty-five hundred more planets?

Poore, Emily. "Kepler Mission Hits 3,500 Candidates." *Sky and Telescope.* November 6, 2013. http://www.skyandtelescope.com/astronomy-news/kepler-mission-hits-3500-candidates/.

NASA, and World Spaceflight News. *Complete Guide to the Kepler Space Telescope Mission and the Search for Habitable Planets and Earth-like Exoplanets: Planet Detection Strategies, Mission History and Accomplishments.*

p. 205   The latest estimate is there may be as many as 40 Billion "habitable, Earth-sized planets in the galaxy"

Overbye, Dennis. "Far-Off Planets Like the Earth Dot the Galaxy." *New York Times,* November 4, 2013.

Mack, Eric. "There May Be More Earth-like Planets than Grains of Sand on All Our Beaches." *CNET.* March 19, 2015. https://www.cnet.com/news/the-milky-way-is-flush-with-habitable-planets-study-says/.

NASA, and World Spaceflight News. *Complete Guide to the Kepler Space Telescope Mission and the Search for Habitable Planets and Earth-like Exoplanets: Planet Detection Strategies, Mission History and Accomplishments.*

p. 206   The Higgs field put a frightening new spin on destiny

Butterworth, Jon. *Most Wanted Particle: The Inside Story of the Hunt for the Higgs, the Heart of the Future of Physics.* New York: The Experiment, 2015.

Baggott, Jim. *Higgs: The Invention and Discovery of the "God Particle."* Oxford: Oxford University Press, 2012.

Close, Frank. *The Infinity Puzzle: Quantum Field Theory and the Hunt for an Orderly Universe.* New York: Basic Books, 2011.

Heilprin, John. "Higgs Boson Discovery Confirmed After Physicists Review Large Hadron Collider Data at CERN." *Huffington Post.* March 14, 2013. https://web.archive.org/web/20130317191649/http:/www.huffingtonpost.com/2013/03/14/higgs-boson-discovery-confirmed-cern-large-hadron-collider_n_2874975.html?icid=maing-grid7%7Cmain5%7Cdl1%7Csec1_lnk2&pLid=283596.

Wilczek, Frank. "The Higgs Boson Explained." *NOVA.* June 28, 2012. http://www.pbs.org/wgbh/nova/blogs/physics/2012/06/the-higgs-boson-explained/.

Zimmerman Jones, Andrew. "What is the Higgs Field?" *ThoughtCo.* January 31, 2016. https://www.thoughtco.com/what-is-the-higgs-field-2699354.

Scanlon, Gerald W. *Higgs Field Unveiled: God's Field at Creation.* CreateSpace Independent Publishing Platform, 2015.

p. 206   According to science writer Dennis Overbye, this puts us in a precarious situation

Overbye, Dennis. "Physicists Anxiously Await New Data on 'God Particle.'" *New York Times*, December 11, 2011.

Overbye, Dennis. "Physicists Find Elusive Particle Seen as Key to Universe." *New York Times*, July 4, 2012.

Scanlon. *Higgs Field Unveiled: God's Field at Creation.* com/?s=Tubular+Solar+Panels+Slash+Costs%2C+Boost+Efficiency.